AF449289

INTERPRETING SEISMIC DATA

INTERPRETING SEISMIC DATA

J.A. COFFEEN

PennWell Books
PennWell Publishing Company
Tulsa, Oklahoma

Copyright © 1984 by
PennWell Publishing Company
1421 South Sheridan Road/P.O. Box 1260
Tulsa, Oklahoma 74101

Library of Congress cataloging in publication data

Coffeen, J.A.
  Interpreting seismic data.

  Includes index.
  1. Seismic prospecting.   2. Petroleum.   3. Gas, Natural.   I. Title.
  TN271.P4C633   1984   622'.159   84-6109
  ISBN 0-87814-259-2

Printed in the United States of America
i

This Book is Dedicated
to
The Indonesians Who Worked with Me
and
My 1983 Class in Athens

My thanks to:
F.T. Barr, who suggested the workbook,
and
The organizations that contributed
materials for illustrations and exercises.

CONTENTS

PART A
START INTERPRETING

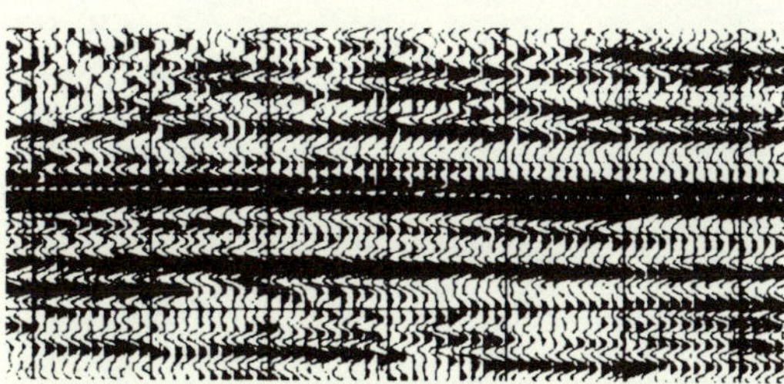

PREFACE

This book, together with its workbook, is a training manual for geophysicists. It is for a beginning interpreter who knows the theory but is not familiar with the day-to-day work in an office. It is for the person brought into the office from working in the field or in a processing center. It is for the student interested in geophysics.

This book is a manual in the sense of explaining how to do things. The workbook will have you actually do some of these things with your hands: interpret a section, tie a loop, contour a map, plan a program, etc.

To interpret successfully, you primarily need to be able to pick seismic sections and contour the results on maps. But there is a lot more that you will find yourself required to do in the office. You will have to communicate with draftsmen about drafting and about reproduction. You will find yourself involved in discussions about assigning lines and about not only whether, but also where, a well should be drilled on your prospect. You will need to be partly a draftsman, file clerk, office manager, financial advisor, geologist. When you interpret, you will need to be able to recognize when you are "spinning your wheels," when you should plough ahead, and when you should start over.

The two books will help you become a little of all these things, with a bit of all these abilities.

WORDS

There are some small matters about wording that should be mentioned at the start of the book.

ILLUSTRATIONS AND FIGURES

In reading this book and doing the exercises in the workbook, you will be looking back and forth from one book to the other. There are sections and maps in both. To avoid confusion between the two, all the diagrammatic and pictorial items in this book are called illustrations, and the similar things in the workbook are called figures. The ones in this book are only intended to illustrate, while the figures are parts of exercises, to work on. When you are reading the text and come to the word "Illustration," look nearby. When you see "Exercise" or "Figure," go to the workbook.

DATA

The word "data" is in an ambiguous situation in the English language. It is somewhere between a singular and a plural. It comes from a Latin word, the plural form of "datum." However, both words have taken on different meanings in general English and in geophysics in particular. Datum is used generally to refer to things that are especially singular, as in "This is a useful datum," or "We corrected to a datum plane." Nobody says "data plane." "Data" is used as a plural, as in "The data are in disagreement," and as a collective singular, as in "The data isn't very good." Most geophysicists wouldn't feel comfortable saying "The data aren't very good." Either "data is" or "data are" can sound correct to the ear of a geophysicist, depending on the context. You'll find them used interchangeably in this book, depending on how they feel in a sentence.

GENDER

There is a gender problem with the English language. Our pronouns are so sex-specific that, to speak of one person of unstated sex, we are often driven to making a deliberate error in number, calling the person "they." I don't want to do that. It sounds as incorrect as it is. I also don't want to make the wording clumsy by constantly saying "he or she" (or "she or he"). Would that English had a pronoun like the Indonesian or Malay "dia," which does mean "he or she." And I don't want to use the traditional "he" for both, relegating the new generation of feminine geophysicists to the category of "Oh, count yourself as included, too."

In this book, things are intended for you, the reader, so "you" will be used consistently. The one concession to the old way is "draftsman" instead of "drafter," which isn't thoroughly accepted yet. Consider "draftsman" to be "draftsm'n," and to refer to either sex.

OIL AND GAS

Like the gender situation, I don't want to clutter the book with the clumsy "oil or gas," "oil and gas," "oil and/or gas," or the academic-sounding "hydrocarbons," so I will just use "oil" in most cases to mean either or both. We usually hunt primarily for oil, but if there isn't any, we are pleased to have gas rather than nothing. Huge gas fields, of course, are welcome, but even then we'd rather have a huge oil field. So I'll usually speak of looking for, or finding, oil, with the understanding that gas is included.

PART **A**

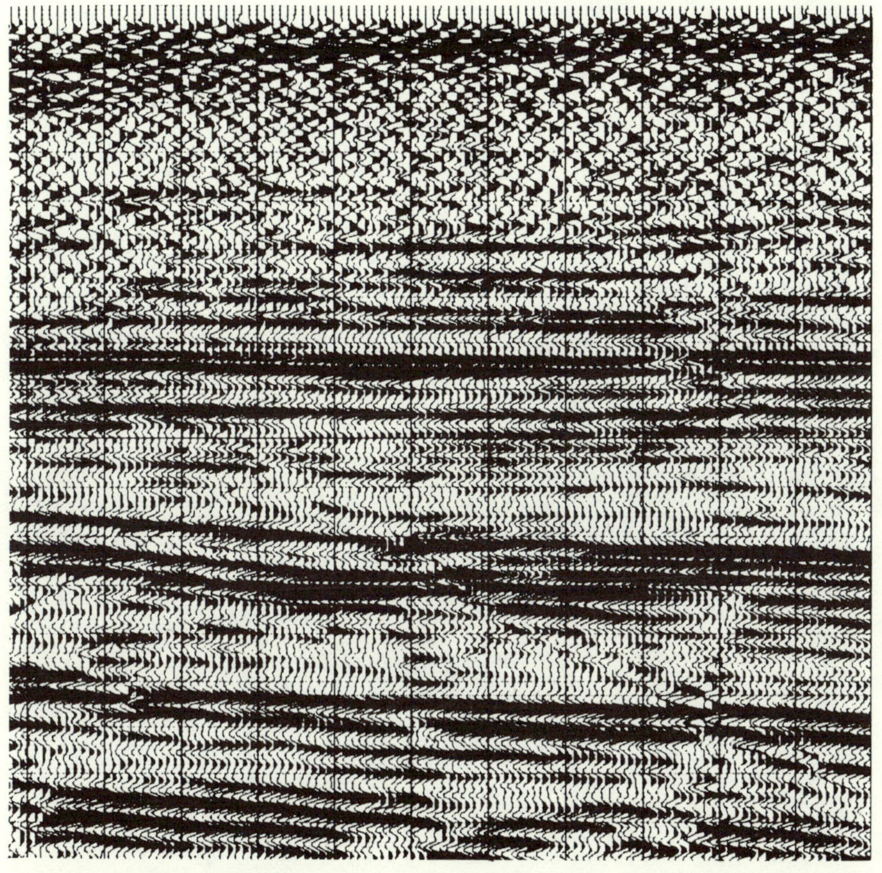

START
INTERPRETING

Once Over Lightly

What is the main thing a person does in performing seismic interpretation? The main activity in terms of both importance and time spent is, of course, picking sections, that is, coloring reflections, the alignments on the sections that represent layers of rock. And the purpose of picking the sections is to learn things about the geology of the area, things that help in selecting locations for drilling wells. But interpretation involves a lot more than just sitting at a desk making marks on sections.

I'll go over some of the types of activity involved in seismic interpreting. One reason for this "Once Over Lightly" is to give a brief summary of the variety of judgments and types of expertise required. Another reason is to put the separate steps into their proper relationships to one another. Then, when we are discussing one topic and need to refer to another that we haven't taken up yet, you will have some familiarity with it. The steps are so interrelated that it is good to have an overall picture of them when we take up any one. For instance, when you work on the timing of reflections, it means more if you know its relationship to mapping and contouring.

When a person is given an area to interpret, some things have already occurred. A company or other organization decided it had an interest in a certain area. Maybe it had leases in the area or only planned to acquire them if the area appeared prospective, or maybe it had a concession or a production-sharing contract. Perhaps the area was already a prospect within a larger area. Whatever the reason for the interest, someone planned a seismic program, normally consisting of a number of straight lines, to cover the area. Then a crew was selected, probably either by hiring a contract company or using the organization's own crew. The planned lines were shot by thumps, pops, shakes, explosions, or some other means. The vibration of the earth caused by the shooting was recorded on magnetic tape.

Those tapes were sent to a data-processing center. The recorded data was processed in a number of steps to make the data better in showing characteristics of the subsurface that are useful in the search for oil and gas. The processed data was played out onto film, which was copied on paper as seismic sections. The locations of the shots were put on a map. The paper sections and shot point location maps are the main raw materials of seismic interpretation.

Now open the workbook to Exercise 1–1, and contemplate an example of these raw materials, the final results of the work of the field crew and the data processors.

We'll wait while you look.

Although seismic sections do come in the form of pieces of paper, one day, for day-to-day use in most interpretation, they will be images on video screens, and we'll indicate our picks to be timed and mapped on an interactive computer terminal. We have been expecting that to happen fairly soon—expecting it since the first use of magnetic recording in seismic exploration, in the 1950s. It is now available and in regular use by some companies, but it may still be some time away from being the normal way of doing seismic interpretation. So we still have these many pieces of paper to handle. Many sections are not only paper, but **big** pieces of paper. So there are things to be learned about just handling all that paper. The sections have to be filed, so they can be found when needed. They should be in a fairly convenient form for daily use.

Interpretation usually starts by identifying alignments on the sections as to what rock layers they were reflected from. Some reflections on the sections are labeled with the names of formations they are believed to represent. A reflection is marked with colored pencil. The coloring is extended along the reflection. Other reflections may be picked also, with different colors to distinguish between them.

At a line intersection, the two lines are matched. The coloring is continued on the new line until an intersection with a third line is reached, and so on around a loop to the starting point. At this point, if the colored reflections on the two lines meet, the loop ties. If they do not, that is, if a pick is at a different time on one section than on the other, an error must be found somewhere on the loop, or a place found where the picks can be changed, to make it tie. In working an area, a number of loops must tie and not contradict one another.

An interpreter must recognize problems on the sections, problems that are inherent in seismic data or that have been introduced in the field or in data processing. There are places where shots had to be skipped, where the line had to be bent, where geophone plants were poor, where field parameters were not the best for the particular problem, where processing differences caused discrepancies between two lines that should tie.

Decisions must be made on extra processing or reprocessing to aid the interpretation. So the interpreter necessarily becomes involved with data processing and, in new shooting programs, determining parameters to be used in the regular processing, and checking the processed sections to see how effective the processing is in solving interpretation problems of the area.

After the loops are tied, the picks are timed by reading reflection times. They are read directly from the timing lines on the sections or by using a scale of time like a ruler, shifted to allow for any distortion of the paper section. These times are written on maps that have the shot points located on them, each horizon on a separate map. The maps are contoured by drawing lines of equal reflection times among them. The contours make it easy to see the locally high areas, which are usually the best drilling prospects. The timing, plotting, and contouring are done either all at once after all loops are tied or in bits to allow the interpreter to see what is happening in the investigation.

Other types of maps may be made and contoured. A time interval map is a map of the differences in reflection time between two horizons. A depth map is a map on which the reflection times have been converted to depths. Usually the maps must be drafted. The interpreter needs to know how to communicate well with draftsmen and needs to be able to do some of the drafting when necessary.

Interpretation differs with the geological conditions in the area. Anticlines are the simplest to interpret, by finding locally high points where rock layers have been bent upward. Faults are interpreted by finding breaks in reflections. The faults can be very difficult to map from line to line. Reefs are built up above the formations they grow on. They are found by recognizing some indication of the buildup. Salt domes are easily recognized, forcing their way through bedding, but their shapes and the positions of traps associated with them are hard to determine. Pinchout and unconformity traps are found by reflection configuration and more subtle clues. Exceptionally strong, weak, or flat reflections on the sections can be clues to the presence of gas in the rock.

Information from different sources may be useful—from written reports, published articles, talking to colleagues, well data. The interpreter should be aware of all the wells in the general area and should know something of the exploration history there.

The interpreter discovers prospects and makes recommendations for developing them further. The recommendations may be for more shooting, other investigations, or drilling. The interpreter presents them to management and, if the prospects seem worth drilling, attempts to persuade people to drill them. The interpreter may write a formal report on the interpretation, to transmit and preserve the information gained in making the interpretation. After a prospect is drilled, the information from the well is available for further search for oil. If a prospect is found to have oil, the development of the field can be guided in part by seismic interpretation.

All that is the work of interpretation, but there are other activities an interpreter is likely to become involved with as a result of seismic expertise and familiarity with the area—dealing with contractors and contracts, supervising field crews, riding boats, climbing mountains. In interpretation and these other duties, a proper exploration attitude is necessary, one that leads to being careful with data but that is optimistic so wells will be drilled.

One
Seismic
Section

Seismic information is obtained by shooting lines, usually straight, which are then processed and displayed in paper seismic sections. We will take up one section alone before going on to relationships between the sections in an area (Illustration 2–1).

THE PHYSICAL APPEARANCE

A seismic section is a picture or diagram of a cross section of the earth, composed of data from individual seismic shots. In its usual form it is made up of many wiggly lines side by side on a sheet of paper. Each wiggly line, or trace, shows a modified version of the vibrations of the ground at a specific point. The wiggles on the trace start at a point near the top of the ground and go on down the paper, representing points farther down into the subsurface. The section is made up of many traces, representing the vibrations of the earth at a line of points on the ground.

Illustration 2–1 Seismic section
(courtesy Petty-Ray Geophysical, Geosource Inc.)

A wiggle on a trace is an indication of the difference between two kinds of underground rock. The rock in sedimentary basins, which are the areas of interest in oil exploration, is usually in the form of layers of wide extent. So a wiggle on one trace is usually accompanied by a similar wiggle on the next trace, and so on, from the same layer of rock. The continuous line of side-by-side wiggles is a seismic horizon, or reflection. This alignment is the evidence on the section of the sound reflected from some layer of rock. Picking a reflection is the process of marking one of these long lines of side-by-side wiggles for further use in the interpretation process.

To make the lineups of wiggles easier to see, some dark areas are usually added to the section by the data processors. The wiggles on the right side of each trace are filled in as variable area, or V-A, added to the wiggle-trace display. The swings to the right on a trace are called peaks, and the swings to the left, troughs. The details of the traces can be seen more clearly in Illustration 2–2, an enlargement of part of the section of Illustration 2–1. The peaks form dark bands, and the troughs form lighter bands between them. These alternating dark and light indications of rock layers cross the section, making it similar to a photograph on a canyon wall, in showing different kinds of rock in layers one on top of another. The section differs from a canyon wall in that, among other things, it does not show depths below the top, but times in which the sound went down to the layers and back up.

The section, then, is primarily made of wiggly traces, with some of the wiggles filled in. In addition, there is some "housekeeping" information on the sections. Distance down the traces represents seismic time. To make these reflection times measurable, horizontal lines cross the section at uniform units of time, typically a hundredth of a second, with the lines at every tenth of a second heavier and at every second still heavier. Some of the heavied lines are labeled at the ends. Using these timing lines, seismic time can be read very precisely, even though the paper may have been stretched or shrunk by age, humidity, use. The horizontal distances on the section are distances, not times.

At the tops of the traces, some of the shot points are numbered. The shot point numbers are also on maps, so a point on a section can be found in its geographical location on a map. Points at which other lines cross the line are also indicated and also points where velocity investigations were made. There may be other points of interest marked on the section, like political or concession boundaries. Gravity or magnetometer data may be shown across the top or bottom of the section. On sections from lines shot on land, elevation of the ground, depth of weathering, shot hole depth (if holes were drilled) may be plotted at

Illustration 2–2 Enlarged part of section
(courtesy Petty-Ray Geophysical, Geosource Inc.)

intervals at the top of the section. Offshore, depth of the sea may be shown. Compass directions of the seismic line are shown by arrows at one or both ends of the section. And a header is on each section.

HEADER

The header of a seismic section is similar to the title block of a map. It is the label that identifies the section and provides details that may be necessary to refer to. A header is normally at the end of the section, extending from top to bottom of the space available. Or it may instead be at the top of the section, extending along it for the distance necessary.

The main items of a header are the line number and area name. Then detailed information follows. The details are usually standardized by one processor at one period, but are not uniform from one to another, and may change from time to time for one processor. Among details commonly included are:

- ☐ Data gathering
    Name of contractor
    Name of client company
    Date of shooting
    Multiplicity of shooting, in fold or percent
    Type of geophone
    Cable length
    Spread layout
    Energy source
    Sample interval recorded
    Recording filter

- ☐ Standard data processing
    Name of data-processing company
    Date of processing
    Degree of stack (may be less than fold shot)
    Sample interval processed (may be less than recorded)
    Deconvolution used before stack
    Deconvolution used after stack
    Time variant filter applied

- ☐ Special processing
    Migration
    Special reduction of multiples
    Coherency filter
    True amplitude display

The headers are often designed to help the processor's people make them up quickly and completely. For this, the header may include a form that lists all the steps that might be used in processing. Then blanks can be filled in about the steps used on a particular section. Most blanks can be completed with numbers indicating the sequence of the processing steps. This simplicity not only saves time but probably makes errors less likely. Examples of contractors' headers are shown in Illustration 2–3.

OTHER LABELING

Some sections have little indication of where the line can be found on a map. This is partly a habit brought on by the need for secrecy from competitors. A competitor with one of your sections, but without the location of the line on the ground, can't use the section to buy leases you may later want.

However, in areas in which companies have large concessions or production-sharing contracts, the need for secrecy is somewhat diminished. Of course, a company doesn't want others to know so much about its business that they have an advantage in buying its stock. Also, data that is too widely known loses value for trading for other data. But there is not much chance for the opposition to see a prospect and snatch it out from under the company. Many oil companies, then, have an index map put on each section. The map usually shows all the seismic lines in the area as drawn lines with line number on each, but no shot point numbers, and with the seismic line of the section itself indicated by a heavier line. This makes it easy to see roughly where in the area that particular line is.

There are some other requirements for a well-labeled section. In all of this, remember that someone will go to the trouble of labeling a section once, but various people will work on it many times through the years. And each time a bit of information has to be dug out, more time is wasted. It is much better to go to some extra effort at first.

On each section, the intersections with other lines are marked to make it easier for interpretations to be made from one line to another. These intersection points are labeled with the line number of the inter-secting line.

Approximate compass directions are indicated with arrows at ends of sections. Most sections have the arrows pointing away from the section. For instance, at the southeast end of the section of a NW–SE line, there is an arrow pointing away from the section, with the label "SE." But be careful, some sections have arrows pointing in toward the section. If an arrow labeled "SE" points down the length of the section, then that label, although correct, is misleading. It is at the northwest end of the line! If you happen to not notice the arrow, you will think the section is going the other way. At sharp bends in a line, direction arrows may point both ways toward or from the bend.

Most sections have all the necessary information in the headers, but in some cases you may not know to look in the "fine print" of the header for some particular piece of information. It has become a rather common practice to indicate special processing or unusual sections by additional, large, prominent labels proclaiming "Wave Equation Migration," "True Amplitude Section," "Special Filtering," whatever may be unusual about the section.

I have been describing how sections normally come from the processor, but they don't necessarily come from a certain processor in a certain set form, like prints from your neighborhood photo shop. Seismic data processing is different. The contracts for the processing involve so much time and money that they justify special arrangements. A processor, for efficiency, normally produces sections in a standard form but is quite willing to display them in any way the client may request.

All the above is what you can expect to find on sections. Now I'll give you some of my own ideas—for use when you may be in a position to suggest the form in which the sections for an area are to be displayed.

To ease the location problem—"this is an interesting lead on Line 265, where is it on the map?"—there are several systems you could use instead of or in addition to the standard index map on the section.

☐   Latitude and longitude, or X and Y coordinates, can be put at each end of the section. This might be useful if for some reason it is not practical to put a map of the sections.

☐   A strip of map with the seismic line down its center can be plotted along the top of the section at the same scale as the horizontal scale of the section. Then, parts of intersecting lines not only appear where they intersect, they also indicate the angle of intersection. Other features of the map—boundaries, coastlines, etc.—are available to help in interpreting the section. Even surface geological features can be included. Any segment of the section contains the map for that segment. The map isn't isolated at the end of the section. This method is probably the most useful. The map contains much information of value to the interpreter, management, whoever may use the section. It also is the most trouble to make.

MIGRATION (24 FOLD)

GEO
NORTHERN OPERATIONS

LINE   SPS
AREA

GEO
SEISMIC SERVICES
NORTHERN OPERATIONS

PROCESSING SEQUENCE

1	DEMULTIPLEX	9	SURFACE CONSISTENT STATICS
		10	CDP RESIDUAL STATICS
1	RESAMPLE	11	RELATIVE AMPLITUDE STACK
1	TRACE BALANCE	12	FINAL T.V. FILTER
3	DATUM STATICS	13	T.V. SCALING
2	TRACE EDITING	14	MIGRATION
4	SORT INTO CDP ORDER		REFLECTION COEF. ESTIMATES
	BAND PASS FILTER		WAVELET APPLICATION
5	DECONVOLUTION		RELATIVE INTERVAL VELOCITY LOG
6	VELOCITY ANALYSIS		ADDITIONAL PROCESSING
7	NORMAL MOVEOUT/MUTE		
8	PRELIMINARY STACK		

DATE PROCESSED   SYSTEM: PHOENIX "1"

DATUM STATICS

DATUM _____ 5600 FT _____   VELOCITY _____ 6500 FT/SEC _____

BAND PASS FILTER

DECONVOLUTION

OPERATOR TYPE _PREDICTIVE_   MIN.PRED.DIST./LENGTH   _28/160 MS_
WINDOW AT MIN./MAX. OFFSETS   _.2-4.0/1.2-5.0 SEC_

MUTE

DISTANCE / MUTE TIME   DISTANCE / MUTE TIME
_990 FT/300 MS_   _7920 FT/1300 MS_

AUTOMATIC STATICS

PROGRAM	WINDOW	STATIC PASSES	PILOT	
ISTAT	.2-4.8 SEC	36	1	
CSTAT	.2-5.0 SEC	36	1	3 TR

FINAL FILTER

TIME	LOW CUT	HIGH CUT	OVERLAP
0.0 SEC	13 HZ	62 HZ	

RECORDING PARAMETERS

RECORDED BY	GEO SEIS SERV	PARTY	19
SHOT INTERVAL	330 FT	DATE RECORDED	
GROUP INTERVAL	165 FT.	INSTRUMENT TYPE	MDS-10
NEAR OFFSET	165 FT	TAPE FORMAT	SEGB
FAR OFFSET	7920 FT	RECORDING FILTER	9/86 HZ
TRACES/RECORD	96	RECORD LENGTH	6 SEC
CONFIGURATION	SPLIT SPREAD	SAMPLE RATE	2 MS
TYPE GEOPHONE	L10A	ENERGY SOURCE	DYNAMITE
GEOPHONE ARRAY	INLINE	CHARGE DEPTH/SIZE	120 FT/30 LBS
GEOPHONES/TR.	18	SWEEP FREQUENCY	
TR. 1 LOCATION	NE OF SP	SWEEP LENGTH	
LINE DIRECTION	NE TO SW	SWEEPS/V.P.	

DISPLAY

_25_ TRACES PER INCH
_2.5_ INCHES PER SEC.   NORMAL POLARITY

BUYER AGREES NOT TO TRADE SELL OR
OTHERWISE DIVULGE THESE DATA TO THIRD
PARTIES WITHOUT THE EXPRESS WRITTEN
CONSENT OF GEO-SEISMIC SERVICES

Illustration 2–3a Section header
(courtesy Geo Seismic Services, Inc.)

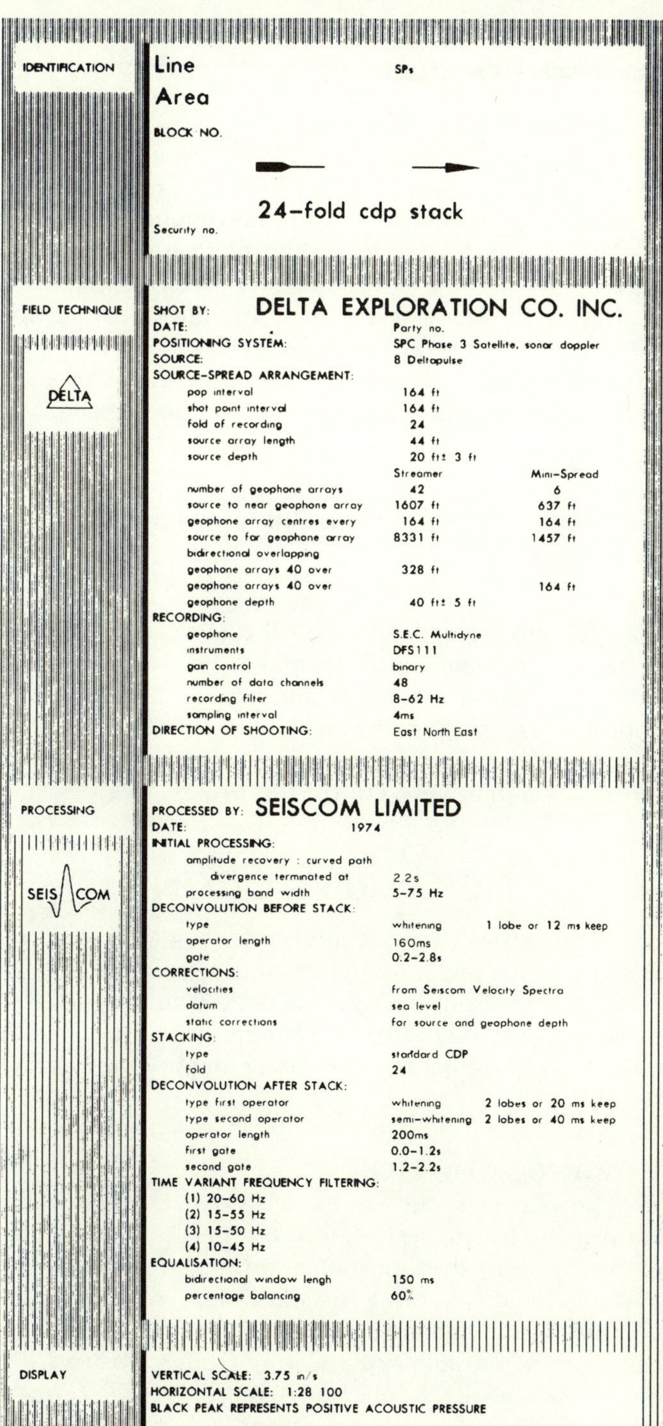

Illustration 2–3b Section header
(courtesy Seiscom Delta United)

In that strip-of-map system, it is difficult to make the map fit the section. In processing a section, and in interpreting, it is convenient to have the traces equal distances apart. But the shot points on the ground may not have been quite uniformly spaced or the line may curve. There are always physical problems in the field. In either case, for the map to fit the section with a shot point on the map directly over that shot point on the section, the map scale would have to be variable.

Some other ways to put map information on a section are:

☐ Just plot a strip of map on the top of the section, and never mind exact fit. A person can look a little way for the shot point on the map.

☐ Plot separate bits of map near the corresponding parts of the section. It may be better to make the map at a smaller scale than the section, with the bits clearly separate. This can be done by just splicing in pieces of an existing map and adding some extra labeling to replace intersecting line numbers and other labels that were cut off the map.

PICK A REFLECTION

The main thing that a seismic interpreter's skill is used for is picking sections. Picking is marking a reflection on a seismic section. It involves deciding what wiggles from trace to trace are from the same reflection, that is, which wiggles were reflected from the same rock layer. In picking a reflection you are determining the configuration of a rock layer in the subsurface.

WHICH WIGGLE DO YOU PICK?

Turn to the seismic section in Exercise 2–1, in the workbook. Look at the section with particular attention to reflections that extend all the way across it. The most evident indication of the reflection is the dark band of the filled-in peaks. But the dark areas overlap from trace to trace, hiding details and making it difficult for marks on them to be seen. People usually mark the lighter-appearing troughs instead. But a geophysicist may pick a peak when it better represents the formation to be interpreted, particularly when there is some need to be very precise about the exact horizon to be picked. Even if you decide to pick a peak, it might be best to do your coloring on an adjacent trough, and remember to time the peak above or below it.

In the normal situation of marking a trough, it is colored in from trace to trace with colored pencil. The marking records the choice you make of what reflection to pick and the decisions you make as to where the reflection goes on its way across the section. The marking also guides you in reading reflection times at the picks, for plotting on maps. A good clear reflection may not require much in the way of decisions, but many reflections that at first appear to be clear-cut turn out to be difficult to actually mark all along the length of the section. There are problems that are not at first apparent, where decisions must be made as to the route to follow in coloring.

Colored pencil is used rather than plain black pencil, both to make the marking stand out prominently and to allow different colors to distinguish between different reflections when several are being picked on the section. In picking, there are likely to be many changes to be made, many times when you will change your mind, so the marking is best done with erasable colored pencil. The different brands of colored pencil vary considerably in erasability. Among most brands, experiment is necessary to find a kind that erases well. However, there are brands that are made especially to erase easily, as indicated by their having built-in erasers. These are the most erasable, although even they do not completely come off the sections. And their built-in erasers are too abrasive for much use on seismic sections. Erasing the same place a few times may remove the image from the section. Soft plastic erasers are gentler on the sections. Unfortunately, these special pencils are not available in many countries.

In the workbook, in Exercise 2–1, pick a reflection on the section.

We pause while you pick the reflection.

Did that? Okay. That is what interpreters spend their working lives doing. There are complications and problems that you haven't encountered yet, but you have made the main kind of decision, where to color in a reflection. Kind of like kindergarten, isn't it? Connect the dots, stay inside the lines in coloring.

If the reflection you chose was one of the more continuous ones on the section, then you probably didn't have any trouble marking it on across the section. You must have noticed that it becomes lower at one end and is a bit irregular. This is because the rock layer it reflected from dips to that direction and is slightly uneven. You can't tell how much the rock dips or how deep it is because on the section you have only reflection times, not depths.

WHY DO GEOPHYSICISTS SQUINT?

Seismic sections are not made with the horizontal and vertical scales equal. Of course not. The horizontal scale is in distance, and the vertical in time. Using the average velocity of sound from top to bottom, the two scales could be made to be approximately the same, but this is not usually done. The sections are more often made with the vertical scale greater than the horizontal. This is done for two reasons.

Many geological features have little vertical relief or are very broad. They would not be likely to be noticed on a 1:1 section. It is more useful to exploration to have the vertical scale somewhat exaggerated with respect to the horizontal. Second, 1:1 scale sections would be longer pieces of paper, leading to greater problems of paper handling and storage. To prevent these problems, seismic sections are somewhat vertically exaggerated. But they must be long enough in the horizontal direction for practical picking. That is, they must be long enough for distinct identification of a point on a section as being directly below a certain shot point.

Even this horizontal scale, though, is too long for easy recognition of some of the subtle lineups on sections. Still more horizontal compression is needed. This can be obtained, in a way, by putting your eye close to the plane of the section, thus viewing it foreshortened. The effect is helped by squinting, to limit your view to the section. This leads to interesting reactions from people who are not geophysicists.

When the section is on your desk, one way to squint along it is to lean back in your chair to bring your head down near the level of the desk and half close your eyes. Another way is to lean forward, laying your head on its side on the desk and closing one eye. One of these positions gets you accused of sleeping on the job about as much as the other does. That's all right, it's still the best immediate way to see what some reflections are doing.

PICK A SECTION

We've discussed the things that can be seen on a single seismic reflection. But a reflection represents only one layer of rock. The geology of the area involves many layers laid down one on top of another and then distorted or broken together, so the geologic picture is composed of all of them together. Some seismic effects also make better sense when considering the entire section, rather than one reflection alone.

Now, do Exercise 2–2 in the workbook, marking several reflections on the section. Pick the best ones, not just because they are easier to pick, but also because they represent the subsurface more correctly than the poorer ones do. Use different colors for the different reflections, if your variety of colored pencils will permit.

Pause.

Got it colored? Now think a bit about the relationships between the horizons. First, you may notice that most of the reflections are reasonably parallel. That makes sense. The rocks are deposited in fairly parallel layers. Also notice that the deeper ones are more sharply folded than the shallower. This sounds geologically reasonable. Tectonic forces distort the deep formations; new layers are deposited on them, flat. The force continues, bending the new layers, but they don't have the first bending in them, so they are not bent as much as the deeper ones. You can look at the spaces between your colored lines to get an idea of thickening and thinning. If the reflections had not been very good, you could have used the knowledge that the reflections are from rocks, which are formed and distorted in certain ways, to help with your picking. If one reflection was wildly different from the others, you might have at least looked at it again to see why it was behaving in a geologically unlikely way. Notice that I didn't say you would necessarily ignore the unusual reflection—if might have been trying to tell you something about the geology. But you should look critically at it.

The interrelationship of horizons is so big a factor in picking that some geophysicists do not like to pick single reflections at all. They feel that they can pick correctly only if they are looking at the geologically broader picture. Others may pick a single horizon, but with an eye on the rest of the section, or may mark a few dips here and there on the section to show them the trends. This latter is my preference. I like to have an idea of what's happening up and down the section, but I also want to concentrate on one horizon.

PHANTOM HORIZON

When a horizon is difficult to pick, other reflections can be used to help you decide how it should be interpreted. An interpretation of a horizon solely by the use of other reflections is a phantom horizon. Where you can't pick your chosen reflection reliably, pick some others, above and below it. They can be short segments of usable reflections, if long ones aren't available. If reflections above and below your reflection conform among themselves, then you can pretty safely assume that your reflection conforms with them, too. You can draw it in, parallel to the trend of these other reflections. Your pick is then a phantom horizon, not a real one that can be seen and marked, but an imaginary reflection drawn in, in a way that you have evidence is reasonable. In some areas the rocks may not be in long continuous layers, so a phantom horizon is the only kind of interpretation that can be made. It shows the trend of the accumulation of lenses of sands and shales, or whatever discontinuous layers there may be.

A phantom can be made by triangle sliding. Place two triangles, set squares, against each other on the section, with another edge of one of them coinciding with one of the visible reflections. Then, keeping the two in contact with each other, hold one firmly against the section while sliding the other against it. By such sliding of one and then the other, thus retaining the dip angle of the reflection, you can transfer that angle to the position of the reflection you want to continue. Draw the angle on the section. Then transfer another visible reflection's dip to your horizon. When you have enough of them at that place, select a sort of average dip to extend your phantom.

Of course, you may not need to go to all that effort. You may, depending on reflection qualities, etc., be able to transfer the overall dip of the visible reflections by eyeball. If so, good. Do it that way.

After making a phantom, you can check yourself by comparing the phantom with horizons above and below it. If, for instance, the phantom gets farther from a horizon above it and also farther from the one below, that probably means that the geologic section is thickening in that direction. But if the phantom gets farther from one and nearer to the other, it is probably incorrect.

Now in Exercise 2–3, pick a phantom horizon.

PICK A LOT OF REFLECTIONS

One interesting and useful type of interpretation of a section can be made by coloring just about all of the section, to squeeze all the geological information you can out of it. To do this, the best reflection on the section is colored first. Having that one colored makes it easier to pick another, as the first one will help you to pick other, weaker, reflections. Going from one to another, a lot more reflections can be picked than it would at first seem. It is useful to mark not only reflections that go across the section, but also ones that go only part way, even quite short ones. The section is now fairly completely interpreted. While doing the picking, it is advisable to color lightly, so marks can be erased as needed. But later, after you have made the decisions in picking, heavying up the colors makes the interpreted geology stand out when the section is viewed from a distance, as when it is put on a wall for a meeting. For this purpose you can go over the coloring by pressing harder with colored pencils or by going over the pencil colors with the wide fiber pens that apply transparent luminous colors to highlight words on a page. The coloring is no longer changeable then, but the horizons really do stand out.

Now pick a section thoroughly in Exercise 2–4.

One form of this sort of complete picking is done by coloring a number of horizons that extend across all or most of the section and filling in the areas between horizons with solid colors. This produces a section on which the interpreted geology stands out even better. If the horizons picked include all the unconformities on the section, the coloring divides the section into meaningful geological units. Depositional and erosional variations in zones show up especially well as thickness changes.

Try this type of picking in Exercise 2–5.

These complete interpretations of sections are mostly for use on one or two lines in an area. The lines you might pick this thoroughly would probably be in your main area of interest, but it would take too much time, for the information you would gain, to do an entire area this way.

IT'S EASY TO PICK ONE LINE

A favorite sport among people who know a little geophysics is picking one seismic line. It's fun for experienced geophysicists, too, including me, but we know more about the limitations involved.

It is nice, and also useful and informative, to see what you can learn from a single seismic line. To start work on an area, you might put a representative section on the wall, look at it, see what features show up on it, color some reflections, look at it some more. This will give you an idea of what horizons you might want to pick on all the lines and of what problems you may expect to encounter in working the area. It might show the types of prospects that might develop in the area. Taking the time to look over a line like this is far superior to just starting to pick a horizon over the area.

That first interpretation of one section is the easiest one. There are no loops to tie, to expose your errors, so you can feel that your interpretation is correct. Later, going around loops, you will fail to tie, and will have to go back and change things, even on that first line. You may have to go over and over some loops, trying one thing and another, before you get a coherent picture of the whole area. The coherent picture isn't necessarily correct, but it doesn't have the glaring errors of mis-ties. It is more nearly correct than an interpretation without that kind of checking.

So what about the non-geophysicist who likes to interpret single lines? A geologist, say, might do it to get an idea of regional depositional and tectonic situations. That is a sensible activity. There is information in the single line. It will be useful in the search for oil. But it may also make the person unaware that there is more to interpreting than coloring one line; it may make the person wonder why geophysicists grub around so long in their interpretations. The answer to the unspoken questions is that, similarly, a person who is good at driving nails may wonder why a carpenter spends so much time measuring and leveling.

It is useful to work with one line alone to get some general impressions, but a thorough interpretation to find a location to drill for oil requires the interpretation of a group of lines together.

Line Meets Line

We have been working with one seismic line alone. But most lines cross other lines. Now we'll take up the interrelationships between lines, starting where two lines intersect. I'll describe how you would handle these relationships in exploration and give you some practice in the workbook.

INTERSECTING SECTIONS

After picking one or more reflections on one line, you will need to extend your interpreting to another line that meets the line you have picked. The place where they intersect is often already indicated on each of the two sections, having been put on the sections in data processing. If not, then you should mark it before doing anything else with the two lines. The company that did the location surveying may have provided a computer printout listing intersections and giving the exact point, like SP 263.46, more precisely than you can spot it on the

23

section. If you do not have such a printout available, then you can go to the shot point map and carefully measure distances between shot points to the intersection. In pencil or anything, plot the point carefully on your section and write above the point the line number of the line that crosses your line there. Do the same on the other line.

Now fold one line back at the intersection point. Be careful to have the fold vertical on the section, perpendicular to a timing line. Lay that fold on the other section at the intersection point on the second section. You now have the sections meeting at their tie point. Match the zero time lines on the two. Check for paper distortion problems by seeing that a timing line down near your pick also matches. If it does not, then shift one section until the timing lines meet, undoing the zero time match. This failure to match is a problem to constantly be on the watch for. Paper sections get damp from humid air; they get hot, cold, dry, rolled tightly. They get grease from arms and the sweat of the brow. All these things distort paper and can cause you to mis-tie if you don't notice. The match that counts is the one right at the reflection being tied. If there is much difference in the two pieces of paper and you are picking a number of reflections, then you may have to keep shifting as you work down, or up, the section.

Look on the unmarked section for the reflection that is marked on the other. It should be at the same reflection time and should look like the other reflection. Mark it in the same color. Similarly correlate any other horizons you are picking. Correlation of seismic sections means matching the reflections and also matching the spacings between the reflections. You have the whole vertical extent of the intersection to look at, so you can compare, not just one reflection at a time, but all the reflections together. For instance, two strong reflections with three weak ones between them form a pattern that should be the same on both sections.

Now you have a start on this new line. You can separate the two, unfold, if the new line is the one you had folded, and continue picking it from its starting point as you did the first line. If the intersection point is somewhere in the line, not at an end, then you can pick both ways from the tie point.

In Exercise 3-1, tie two lines where they intersect and mark the horizon from one line onto the other.

That's that, except for the possible problems. We've taken up one of them, the possibility of paper distortion. There are others.

The two sections may not be the same scale at all. There are a number of conventional scales for the time scale on the section. One second may take up 5 centimeters, or 10, or 3.75 inches, or 7.5, or some other distance. It is common to have sections produced at one scale and then photographically reduced to half that scale, and you might somehow wind up with one full scale and one half scale, if some sections got lost or misplaced. If your sections are at such radically different scales, you obviously can't match them as I described. The most elegant solution is to request, and get, sections that match. But elegant solutions have a way of eluding geophysicists. Next best is to have one of the sections photographically reduced or enlarged to the scale of the other. This too may be impractical because of a deadline or some other reason. You are likely to have to live with the two scales. What do you do?

The quickest way to handle this problem is to look on one section for a reflection at the same reflection time as the one on the other section, and see if they look alike. This isn't very visual. It doesn't let you compare the two sections all up and down, and so isn't nearly as reliable. But it is necessary many times in interpretation. Of course, the way to look at the same time on the other section is to read the time of the reflection, at the intersection, on the first section; then locate that same time at the tie point on the other section; see if there is a similar-looking reflection there. This same technique can be applied to other situations, for instance when trying to make a correlation by telephone. Read off the time at the intersection to the person who has the other section, who can look at that time for a reflection. Then you can try to describe the reflection to each other. Not very satisfactory, but it could be necessary between, say, an interpreter and a processor.

We can use this same technique as an exercise now, in Exercise 3–2.

There is a way to make the match between sections of different scales more visual. Fold the larger-scale section and lay it at an angle on the smaller. Choose the angle so two timing lines on each section match at the fold, for instance, the one-second and the three-second timing lines. Then all the reflections meet at proportional distances, except for one factor. With that angle, the sections are now meeting at the intersection point only at one seismic time. So it is correct only if the reflections on the small scale section are perfectly flat for all the distance the other section is spread over. So it works fairly well if the reflections are fairly flat. And it works fairly well if the two scales are not very different, so the angle isn't great.

Whatever the relative scales, the same or different, when you make the timing lines on the sections match, the reflections may not fit. Sometimes the reflections will match only if you make the timing lines miss each other by some fixed amount. What does that mean? Well, the reflections must match. The intersection is a point, and the reflections represent the same rock layer. At one point, the rocks can't be at one depth on one section but at a different depth on the other. The problem has to be in the sections.

It may be that the sections were processed differently. This is probable if they are from different batches of shooting and processing and is possible if they are from the same batch. It is hard to get a grip on this from only one intersection, though. We will go into the problem more thoroughly when we are considering a number of lines together.

Then there is another problem. Suppose the reflections don't match even when you shift the sections. That is, you shift the folded section up and down on the other and can't find any position in which they fit. Maybe two recognizable reflections are farther apart on one section than on the other, so no amount of shifting up and down can possibly make them fit. The first and most likely explanation is that someone—you or whoever else marked the intersections on the sections—made a mistake. Check in every way you can before looking for another explanation. Look the line up on the shot point map and on the printout. Either might be wrong, somehow. Check the survey notes if they are available.

Another possibility—unlikely, but it has happened—is that a whole seismic section is mislabeled, with the shot points listed in the reverse order, or all shifted by some uniform amount. The section could even have the wrong line number on it.

Then, after exhausting these possibilities, you can consider that there may be a surveying error. If the sections are from two different shooting projects, one project as a whole might be shifted with respect to the other. Or within one program there might be a localized survey error.

See if you can determine the intersection in Exercise 3–3.

While we're on the subject of line intersections, I'll make some comments on the labeling of intersection points on the sections.

Where two lines intersect, the point of intersection is normally marked on each section by the data processors—if both lines are in a group of lines that were processed together. The processors won't necessarily know about lines that were shot and processed earlier. The client company sometimes provides processors with a base map of earlier lines so the intersections with those lines can be added to the new sections. And lines shot later obviously couldn't have been taken care of in the current processing. So there are usually some intersections that are not indicated on the sections you have for interpretation. But these markings are essential to your interpretation, and so have to be on your sections. You have three choices: mark intersections on the sections one by one as you work the sections; put the intersections on all at once, in a big chore before you begin interpreting; or, best of all—if you have the foresight and the time, and the draftsmen are free—get the draftsmen to ink intersections on the films of the sections before your prints are made.

Here is a suggestion for labeling intersections that may make it easier for you to keep track of which sections are being tied. The standard way for intersections to be marked is to put the line number and shot point number from the intersecting line on your section, with a pointer to the exact spot. The top of a section with two intersections might look like this:

Line	203B	A82R
Shot Point	1634	59
	I	I

This is nice. But, in interpreting, one section is folded at the intersection and placed over the other at the intersection point. You don't then have the line numbers at the ends of the sections in view. On each section, you see the line number of the other section. It's a little extra mental effort, just when your mind is concerned with a number of factors about correlation, ties, and other interpretation problems. I like to have the line numbers of both lines and the shot point numbers of both lines marked at each intersection, like this:

Tie Line	203B	A82R
Shot Point	1634	59
This Line	12	12
Shot Point	422	486
	I	I

Now, when a section is folded at an intersection, the mental gymnastics are reduced. The Line 203B–Line 12 label can be matched with the Line 12–Line 203B label and Line A82R–Line 12 matched with Line 12–Line A82R—easy, same numbers—and you can look whenever you need, to see which line you are working on.

An alternative is to have the line number of the section repeated frequently along the top of the section.

Line 12	Line 12		Line 12	Line 12
Line		203B		A82R
Shot Point		1634		59
		I		I

That takes care of some details of labeling. The interpreting, which is the main thing you are doing with the sections, is made easier by good labeling.

TIE A LOOP

Lines are usually shot so as to form loops. After one intersection is tied, you can go on to tie the other intersections. This lets you keep on until you meet yourself back at the first intersection. That allows you to check your interpreting. When you get around the loop, what if you are not on the same horizon? You've done something wrong, that's what. Wrong, but not necessarily blameworthy. In geophysical interpretation, everybody mis-ties loops. If there could be a tally of all the times people have worked their ways around loops, I'm sure the mis-ties would turn out to outnumber the ties. Of course, one reason for this is that when you mis-tie, you try again, and keep on trying until you tie. But then you normally stop. That tie isn't necessarily correct either, but you at least don't have evidence that it is wrong. A tie doesn't show that you are right, but a mis-tie does show that you are wrong.

If, on the first time around, you fail to come back to the same reflection you started on, what do you do? First, look for a plain mistake, like a place where your pick wandered off the reflection onto another one. Or, if you had sections that were at different scales so they couldn't be directly compared, see if you made a numerical error in going from one to another. You might check to see that you used the ties correctly and didn't match the right place on one section to the wrong place on the other or tie it to the wrong other section.

If you fail to find the problem with that investigation, look over the sections again, particularly at any places where you had to make decisions. In the light of knowing that you didn't tie, you may decide that one of the decisions should be made differently. If that doesn't solve your problem, you may just have to look for the poorest reflection quality on your horizon and force a change there. That doesn't sound very scientific, but neither does a mis-tie. If the loop doesn't tie, something has to give. The two lines at an intersection are at the same place, so they must agree there. Disagreement of the rocks at the intersection is physically impossible. Some other place, within one line, there can be an alternative way of picking that, although it may be hard to recognize, is not impossible. Make any changes in picks between the intersections; the intersections themselves must tie.

CONTINUITY AND CHARACTER

There are two main ways of picking reflections, by continuity and by character. Without calling it by name, we have been stressing continuity, the continuous appearance of a seismic reflection. A strong reflection looks like a path of light or dark across the section, or rather a combination of light and dark bands. A poorer reflection may not be so visibly continuous but may also be pickable by just marking it continuously across the section.

Character is the shape of a reflection, the characteristic look of the combination of wiggles of a trace that make up the whole reflection. This could be, for instance, two strong peaks with a strong trough between them. Or it might be a strong trough, two small peaks together, and a weak trough. The words peak and trough derive from the time before seismic sections, when individual paper records were interpreted. Handling several of those long strips at a time was easiest if they were laid across a desk from side to side of it. The end representing the surface of the earth was customarily placed to the interpreter's left, with reflection time increasing to the right. From this vantage point, an upward wiggle was called a peak, and a downward wiggle, a trough. Or if you don't want all that history, just think of the way a single trace looks on a section that you squint down, with the top of the section to your left. A wiggle that has a smaller wiggle in the opposite direction, within it, making it look like two smaller wiggles, is a doublet, or a saddle.

When continuity fails you (because for some distance along the section there is very poor reflection quality or a fault) or when you need to correlate two sections that do not meet, then character takes over. If the characters are similar enough on both sides of the break in continuity, then you can decide that the two events are the same reflection. You also use the characters of other reflections and their distances apart, when you can, to correlate the section, rather than correlating just one reflection alone. But there are times when the character of one reflection alone must be correlated. If a reflection is by far the most recognizable one on the section, then it may be the one that guides the correlation of the others. Or if thrust faulting, reef growth, or something makes the one reflection wander about independently of the others, then it must be correlated alone.

If continuity is good, you can quickly and easily use it alone to pick a reflection. But when continuity is good, character is usually also good—and is also influencing you. The good, continuous trough may be between a strong peak and a lesser one, so you recognize this sequence as well as the fact that the trough continues. But when reflection quality is poorer, you may need to use both continuity and character to get the reflection picked. And when the two seem to disagree, then you have a problem.

Beginning interpreters often ask which to believe, character or continuity. There isn't any general answer. The two are normally expected to go together. Sometimes an undetected fault allows two different reflections to meet so well that the continuity looks quite good. Then a person might pick straight across, and so might mis-tie. Close examination and the need to change something, somewhere, might make the interpreter notice that the character of the reflection changes abruptly at that point. In another situation, gas trapped in the rock, or coal within the formation, or a change in the thickness of the bed may produce a character change, but the correct interpretation might be made by following the continuity.

Seismic sections are designed primarily to show reflection continuity. Character correlation from place to place might be accomplished better by short pieces of sections. A short piece of section could be readily picked up and laid over another to compare reflections.

In working a seismic section, geophysicists normally just look along it and, to jump over a fault, poor record zone, or gap in the shooting, look for similar-appearing reflections. A better correlation is achieved by folding the section back on itself and laying the fold over the other part of the section. This is an activity that has become characteristic of geophysicists: making a fold in a section, then moving the folded edge forward to another part of the section, and looking up and down the section to see if the reflections correlate. The folded edge is shifted up or down to find the best correlation. Then the reflection that was marked on one side of the problem is marked on the other side.

Correlate by folding, Exercise 3–4.

WHY DIDN'T IT TIE?

There are many problems that make the interpreter's job difficult, but if it wasn't hard to do, companies wouldn't need us, would they? Loops fail to tie, and we have to go back over them, looking for something to change.

The easiest sections to interpret have good, clear reflections and little or no dip, but they aren't likely to lead to oil discoveries. Anomalies of the kinds that contain oil, because of dip or hydrocarbon content, may be difficult to pick and to tie.

We have already taken up the possibilities of mislabeled lines and intersections, and have touched on the need to force a change if reflection quality is poor. Other problems include character changes, reflections that split in two, faults, survey problems, processing problems, even the very anomalies we are looking for.

Even starting with a good reflection, its quality may diminish farther along the section, and then it is not the good, strong reflection that it is at the starting point. This happens for a variety of reasons, both geological and geophysical. When reflections are poor, either locally or all over the area, the interpretation is difficult and its results unsure. There is not much that can be done about that in interpretation. It is a problem that, if it can be solved after the shooting is done, must be solved by processing.

When character changes cause mis-ties, the solution is more likely to be in interpretation. Careful correlation from place to place and comparison with the continuity may allow you to locate the place where the change takes place. Then you may be able to account for it geologically, say by a pinchout or lithologic change.

These and other problems may cause you to mis-tie. They can be handled better by considering a number of intersections, so most of them will be taken up later, as problems in working an entire area.

TRICKS OF THE TRADE

The purpose of this book is to help you become a good interpreter in all respects. You have started picking sections, and we will cover other major aspects of interpreting. But there are also some "tricks of the trade," little things that people just think up as they go along, that I will put in here and there as they apply. Here are some of them.

When you are trying to correlate across a bad spot on a section and find that folding can't move parts of the section to the right position without tearing the paper, there are other ways to handle the correlation. It can be done more efficiently by:

1. Cutting the print in two. Then the parts can be overlaid, folded, overlaid differently, as much as is needed. A little tape on the back—not the front—applied with some care in matching, makes it as good as new—or nearly so, depending on the match.

2. Having two prints of the same section. Then one can be folded and matched to the other.

3. Making a xerographic copy of part of the section. It may not be a very good copy, but it is quick and expedient. Fold it and lay it over the section.

Faults are often found as breaks in the continuity of reflections, and people must correlate across them by guess and by eyeball. The correlation can be made orders of magnitude more reliable by a simple, quick method—cut the section apart along the fault plane. My wife and I discovered this in one interpreting project. We were then madly chopping up sections, temporarily putting them back together with a couple of bits of tape—in case we might want to recorrelate—and handling mighty lacy sections. In Exercise 3–5, correlate across some faults that have been cut and rematched for you.

In the case of a growth fault—a fault with throw diminishing upward—the two pieces can be matched at the surface, with little or no offset, then shifted to match a slightly deeper reflection. Shift again, and so on, working your way down. Some amazingly good correlations can be seen this way.

I feel that this technique, which is so simple that it is probably used by a number of people, but so little discussed that we had to find it for ourselves, is equivalent to quite a large advance in data-processing technology. And all it costs is a little more frequent replacement of section prints—and maybe some extra pairs of scissors around the office.

This correlation of faults, too, could sometimes be done by folding another print, but most faults aren't straight, and it's hard to fold on a curved line. Of course, it can be done by making a copy, cutting it in two, and laying one piece over the original section.

These folding and cutting techniques are used in interpreting by hand. A computer interactive interpretation system allows you to handle the sections in similar ways, all on the screen. If you have one available, it's probably a faster way to handle the data and is also better in some respects. There isn't any paper stretch, and there aren't poor prints. On the other hand, computer terminals are dependent on a steady source of power and must be in good working order. Neither of these conditions is certain to exist whenever you need it. And you may—very likely will—have some sections for which you do not have the recorded tapes and so can't put them on the computer.

Besides, I can't fit an interactive terminal into this book or the workbook. And if you learn to do the work on paper, it will only be easier when you later have a terminal available to work on.

REVERSED SECTIONS

When you are trying to tie a difficult loop, you may feel that you could locate the spot causing the problem better if you could see the whole loop all at once. Loops are made up of parts of lines. When the lines are shot in a rectangular grid, a loop is parts of four lines. For convenience, let's discuss the rectangular grids, recognizing that the situation is similar for other loops, whether they are triangular, one curved line meeting itself, etc. To pick a horizon around a loop, you usually fold a section, lay it over another section, mark your pick from one to the other, then pick that section to another intersection point, make another fold to go to a third section, and so on. By the time you have been across all four sections and have found you didn't tie, it's hard to remember what all the sections look like. Maybe they all look fairly good. But there's an error somewhere, or the loop would tie.

How about cutting just those parts out of the four sections and laying them end to end? Fine, but when you try it, they don't fit together. Two are backward to the other two. Naturally. East-west sections are made with the east end to the right, so two of the lines have their east ends to the right. But to unwind the loop, one of those needs to have its east end to the left. You find you can put two lines together and also the other two, but not all four. You get to look at more of the loop at once, but not all of it.

If there is time, you might get the data processors to play two of the lines with their traces in the opposite order, or you might even get them to make a section of the loop, played out as one line. But this means waiting, not solving the problem while you have it well in mind, but having to set it aside and pick up where you left off, some time later. It also costs money, and it can be subject to error because of misunderstanding between you and the processors.

There is a quicker, but not so elegant, way to see the loop as one line by looking at parts of it backward. This means that the backward part has the peaks and troughs reversed, but the filled-in variable-area dark parts will line up properly. There are several ways to do this.

1. Lay the pieces of sections on a light table, with some of them turned face down and the pieces placed, or taped, together at the intersections. Have the light shine through them and work on whichever side is up.

2. Get reversed diazo prints of some. This calls for making the prints with the transparent master placed face down on the diazo material, that is, turned over from the way it is in normal printing.

3. Use transparent copies of all the lines and turn some wrong side up.

4. Make reversed prints of some on the office copier and tape them and normal prints of the others together into one section.

5. Use an interactive terminal to reverse some lines and to display the different pieces of lines matched up as one line.

Number four is usually the most practical. The way to make the backward copies is to make transparent copies on the copier, place them wrong side up, and make normal opaque paper copies of them. Special film is sold for making transparent copies, mostly for use on overhead projectors. It is inexpensive and is a good thing to have on hand. Plain drafting film or tracing paper also works in copiers. Just cut it to the right size and load it in the copier. Don't use very thin paper, though, like onionskin. It isn't stiff enough to be moved forward by the rollers and would wad up and jam the machine.

However it is done, the sides of the loop are together as one line. It is also convenient to have an additional copy of one end added to the opposite end. The part that is repeated should be a part of a line with good record quality, a part that is not likely to be where the problem is.

With the end repeated you can see how things tie without any folding or bending. Mark the pick at one end of the section and at the same spot on the other end. Now you have one section on which you have to pick a reflection from one of those marks to the other. You need to indicate the intersection points by marks on the section, so you will be aware of dip changes to be expected at those right-angle bends. The problem of the tie may still be difficult, but you at least have it all laid out so your eyes can go from one weak point to another for comparison, rather than looking at one section and trying to compare it by memory with another. Like all the other tricks, this isn't a cure-all, but it may sometimes help you do better interpretation. Work Exercise 3–6, picking one of these put-together loops.

Another use of this same technique is in putting together parts of lines to assemble a section that connects one well with another. If one line crosses one well and another crosses the other, then there may be a series of lines, perhaps zigzagging, that connects the wells. Prints of the segments of lines can be spliced together to make one continuous, even though crooked, section from well to well. To assemble this section, it may be necessary to reverse some of the sections. As in putting a loop together, bends should be indicated on the composite line.

These techniques will help you pick reflections and tie loops. This brings us to the stage of working with an area. We'll start with identifying reflections in terms of geological formations.

PART **B**

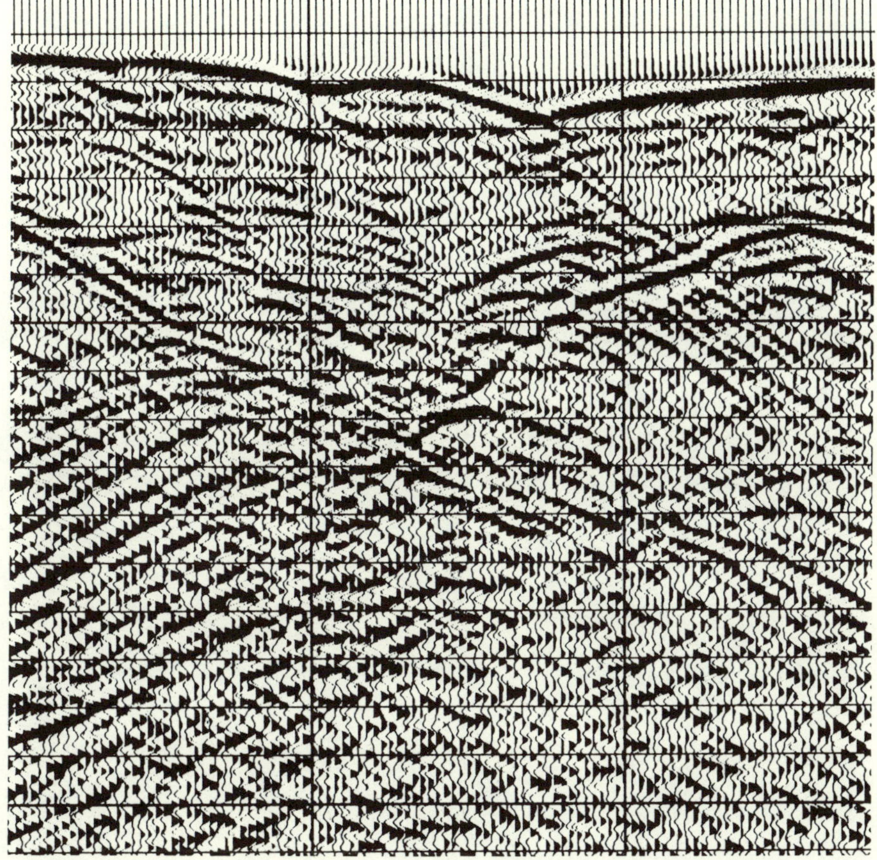

AN AREA
AS A WHOLE

Identify a Reflection

In the interpretation we have been doing, any necessary identification of horizons had already been done. But interpretation of an area for which identifications are not already made requires that you identify seismic horizons as to the formations they are reflected from. An experienced interpreter usually starts working an area by identifying horizons to be picked.

We will first discuss some of the characteristics of the wiggles we will identify; then we will take up specific ways of making the identifications.

WAVELETS AND PHASE

Seismic sections are made up of traces. The traces are made up of wiggles. The wiggles represent reflections. The wiggles can be displayed in either minimum-phase or zero-phase form.

WAVELETS

The wiggles on a seismic trace originate from the shot or other source of energy, reflected back upward from many layers in the earth. In analyzing the wiggles, the trace can be broken down into wavelets. A wavelet is the little group of a few wiggles that was put into the earth by the source and that was somewhat changed by its travel through the earth. The trace is a combination of these reflected wavelets—some strong, some weak, some wiggling one way, others in the reverse direction, some overlapping and combining into different shapes. The form in which the wavelet is displayed has an effect on which wiggle is best to select and pick to represent a certain formation.

MINIMUM PHASE

In a minimum-phase wavelet the energy is spread out in time, with the strongest energy in the early part of the wavelet. This is more or less to be expected in any effect that starts suddenly—that it will diminish with time. The reflection from one velocity interface is a series of wiggles, for example, a strong peak and trough and three or four weaker ones of each. To truly pick the reflection's response to the interface in the subsurface, you would need to pick the very start of energy building up to the first noticeable peak or trough. That start isn't easy to detect, though. And picking the exact start isn't all that necessary. Another part of the wavelet will continue across the section at some fairly uniform time after the start, so a simple subtraction will adjust the time picked to the initiation of the reflection. It is more important to pick an event that can be reliably followed across the section through the obstacles of poor reflection quality, faults, lithologic changes. Your picking will be more reliable if you choose a strong, that is, high-amplitude, part of the reflection.

Your pick should also be one of the first parts of the reflection, nearest the top of the section. There is less adjustment needed to bring it up to the time of the reflecting surface to tie wells and therefore less chance of error caused by interference from other reflections or changes in this reflection. So in most of your interpretation you will probably pick a strong trough, early in the band of energy from the reflection.

ZERO PHASE

A zero-phase wavelet is composed of wiggles that are symmetrical around the reflection time of the horizon. The wiggles of a reflection build up from small wiggles to the largest at the reflection time and then taper off in a pattern that is a mirror image of the part above the reflection point. This type is best picked at the strongest wiggle, the highest amplitude. Of course, energy can't arrive at the surface from a rock layer before the reflection time of sound traveling down to the layer and back up. This zero-phase situation is created from the data after it was recorded, putting half the energy above the reflection point. Most vibrator data is assembled as zero-phase sections. Also, many other sections made by wavelet processing are played out as zero phase.

If a section is minimum phase, then for a band of energy representing one horizon, you should pick the first strong energy in that band. If the section is zero phase, the nearest energy to the reflecting horizon is the strong central peak or trough. That peak or trough is, within the abilities of the system, exactly at the reflecting horizon. Whether it is a peak or a trough depends on the polarity of the section and the polarity of the reflection.

Seismic data may not be in exactly one of the two forms, minimum phase or zero phase. This can affect character and time of reflections. For this and other vagaries of seismic data, a little adjustment of data may sometimes be needed to make identification of horizons work well.

WAYS TO IDENTIFY

Formations cannot be identified directly from seismic sections, in the present state of geophysical development. They are originally identified, named, even defined as being formations, in geological work, from outcrops, well logs, or both. A geologist may see an interesting outcrop, follow it by its appearance and fossil content over the countryside, and give it a name. Later that same geologist or others may recognize it by seeing the same fossils in cuttings from wells, and may then extend the recognition on the basis of distinguishing features on well logs.

After that has been done, the formation can be correlated from well to well. If there is a seismic line that passes close by a well, the well logs can be compared with the section at that point. But the points on the well logs are measured in depth, as so many meters or feet below the rig floor. Points on a seismic section are measured in seismic time as so many seconds below (after) the zero time on the section, which is at some arbitrary datum plane. These aren't the same units, and they aren't measured from the same point. There has to be some conversion made to fit the two together. Velocity ties together time and depth.

Velocity of sound varies greatly from place to place and with depth. We can relate a seismic section to a well if we have a set of measurements of the velocity of sound at the different depths in the well. Those velocities, with a correction for the different starting points of well data and seismic data, would make us able to associate formations encountered in the well with reflections on the seismic section.

Some wells have had surveys run in them that provide artificial seismic traces (synthetic seismograms) or short sections (vertical seismic profiles, or VSPs) that can be correlated with the seismic data. In the absence of well data, there may be seismic sections, already identified, to correlate with the unidentified sections, or other clues to the identification of reflections.

There are a number of ways of making the identification, with varying degrees of reliability, depending on the amount and type of information available. We'll take these up, not starting with either the best or the worst, but in a sequence designed to make the processes more understandable.

ALREADY IDENTIFIED

The simplest method is the use of your or someone else's experience, as when someone has marked the identifications on a section. Up to this point, when we have used reflection identifications, we have been using this method. You can also identify reflections by a tie between the section and an already picked section in another area. You may have previously worked the other area and may be extending your interpretation from it into the present area. Or another interpreter may have worked the other area. In either of these cases, it may be necessary to have the identifications agree from area to area so the interpretations can be joined together. Even if you think the first identifications were not exactly right, you may decide to continue picking those same horizons, with maybe a note added to the sections and maps to point out the disagreement.

FAMILIARITY

Whether there are wells or not, the easiest identification is the same type you use to identify a friend's face—by already being so familiar with it that you recognize it at a glance. This occurs with seismic data in areas you have worked over a long period, in which the reflection has a distinctive appearance. You may instantly recognize a very strong reflection, or a reflection with a characteristic double peak or something, or the only good reflection on a section. In northern Alberta, for instance, experienced interpreters recognize the strong, shallow, characteristically shaped Wabamun reflection so readily that they might use it as a reference point on a section, even over the telephone, just as they might refer to the two-second timing line.

TIME-DEPTH CHART

A time-depth chart is a plot of seismic times against depths. In one form it is a graph on grid paper with a curving line on it. You can look up a time on it and read the corresponding depth or look up a depth and read the time. In another form, it is a tabulation, a list of times with a depth given for each time. The graph is quickly made and is convenient for looking at the overall velocity distribution. The table takes more effort to make, gives more precise readings, and can be read more quickly. If you have a list of formation tops made from a geologist's picks on a well log, you can use either form of time-depth chart to read the times at which there should be reflections on sections.

Now do Exercise 4–1 to get some practice with this process.

STACKING VELOCITIES

The velocity information obtained in processing the seismic data can be used in identification. This is the seismically derived velocities, that is, the stacking velocities and the average velocities derived from them. These velocities are obtained by the processors at intervals along the section. They are fairly reliable in the shallow part of the section, pretty unreliable in the deeper parts. If you look at these velocities carefully, you will see that they can change erratically from one velocity determination to the next. The true velocities are not that variable, but the uncertainty of picking stacking velocities makes them vary widely.

Rather than use the velocities from one velocity analysis alone, it is better to average them from several nearby points, preferably including some from more than one seismic line. Then use that velocity function to match whatever formation depth information is available, from a well or any other source. You can make a time-depth chart from this velocity information. But since velocity analyses aren't very accurate, you should consider the time-depth relationships as approximate. You need to apply judgment about reflections and formations to help you decide which reflection is from which formation.

A large velocity difference ought to produce a strong reflection. If you do not have velocity information from the well, you have to decide on the velocity differences from a general knowledge of velocities of different lithologies. Shale is very compressible, so it has a slow velocity if it is shallow, but the velocity increases with depth. Limestone, salt, granite have high velocities and are not very compressible, so their velocities are largely independent of depth. The velocity of sandstone depends very much on what material fills the spaces between the sand grains.

Now work Exercise 4–2, using the stacking velocities to help identify reflections.

SONIC LOG

The identification situation is better if there is a sonic log in a well in the area, a log that was run for most of the vertical extent of the well. Some sonic logs are run just to determine porosity in geologically interesting zones. These logs are quite short and are quite useless for determining overall velocities. Only fairly long sonic logs are useful for seismic velocity determination.

A sonic log is a log of instantaneous velocities. It is continous, so the information forms a continuous line. There are many little changes in velocity—at all the tiny layerings in the sediments. This makes the line very irregular, jumping back and forth at a high frequency. It shows many little changes of velocity but does not show average velocities for thick intervals. These average velocities are necessary to calculate depths to reflections.

Some logs are integrated by data processors. That is, the velocities in them are used by a computer to calculate the times represented by the depths in the well and plot the ten-millisecond intervals as tick marks on the log. The tick marks are closer together where velocities are faster, and farther apart in the slower zones.

If your log is not integrated and you don't have the time or budget to have it done, you can perform a crude integration by hand. On the log, draw vertical lines at average positions of the sonic curve, through the middle of its little jumps back and forth. At a large change of velocity, start another vertical line. At the ends of your vertical lines, read the depth and obtain the velocity. The log doesn't give velocity but does give the reciprocal of velocity. You need to convert it to velocity by dividing it into the number one. Then divide depth by velocity to give the time. Multiply that time by two to get two-way time, like the time on a seismic section. Mark that doubled time on the log.

But sonic logs do not start at the upper ends of the wells. They do not work correctly in cased holes. They would tend to measure the velocity of sound in the steel casing. So a sonic log is started below the surface casing. This leaves a gap above the sonic log with no velocity information. The times and depths are loose, to be slid up or down at will. What do you do about that?

If there is a good reflection that you recognize as being from a certain formation, you can just identify that reflection as being from that formation, and the other parts will take care of themselves. But if this isn't the case, and it often isn't, there is another trick available.

Plot the times vs. depths on grid paper—the ones you obtained from averaging velocities or some from an integrated log. If you use points from an integrated log, you don't need to use a lot of points. Just use enough of them to define the overall curve, including points at large changes in velocity. The plot started at the upper end of the sonic log, which is below the surface casing, not at the starting time of the section, and the depths are from the KB, kelly bushing. You can plot the level of the top of the section on your paper, up above the start of the log. The gap you need to fill in is from that point down to the top of the curve you have drawn. You need a zero time on the chart at the zero time of the section.

You should find some shallow velocity data to incorporate in the plot. Most modern sections have stacking velocities noted at the top of the section. Stacking velocities lose accuracy with depth and are better in the shallow zones. That's fine, the shallow part is what you need. And you really need average velocities, that is, velocities that are averages of all those down to the reflections, rather than the stacking velocities. Average velocities can be calculated from the stacking velocities and are often noted at the top of the section. Use the best velocities you have. Plot them downward from the datum plane on another sheet of grid paper. Join the two curves and trace the first one onto this new sheet. You now have a time-depth chart that can be used for identification and also for converting times to depths.

If you don't have any shallow velocity data, you can extend the velocity curve you have plotted. Project it upward at the curvature it has, to the elevation of the datum plane. Then trace the curve onto another sheet of grid paper with this top point at zero time and zero depth.

VELOCITY SURVEY

More accurate than just a sonic log is a velocity survey. It is made by shooting a velocity survey with check shots and combining the results with the sonic log velocity information. The sonic log itself provides good, detailed information on velocity but is subject to cumulative error, making the overall velocity incorrect by about ten percent. So the velocity is more accurate if the log has been adjusted to fit a velocity survey. The idea of the check shots is to determine velocity that will apply to a seismic section, by using a technique similar to that used in making the sections, shooting and recording. The shots, as in ordinary shooting, are fired near the surface of the ground. But the recording is made from a geophone at known depths in the well. Thus, travel time from the surface down to the formations is measured. Twice those times should be the reflection times—down to the horizons and back to the surface.

A specialized geophone is lowered to the bottom of the well. A seismic source is fired at or near the surface. The source is often a marine-type air gun placed in the mud pit of the well. A shot is fired and recorded. The geophone is pulled up some known distance, another shot fired, and so on. The recording is done on the way up, rather than down, to make sure that the depth readings are not confused by the geophone's getting stuck in the hole and letting the cable hang slack.

There is no special need to have the geophone at specific horizons. It is sufficient to record with the geophone at even intervals in the hole, as the sonic log will provide the detail about formation changes. Each shot is recorded at several gain settings to obtain different amplitudes. The first break (the first energy to be recorded) is read from whichever amplitude happens to indicate it the most clearly. Its time is the **one**-way time from shot to geophone for that depth.

These times to the depths are considered correct and are used to calculate velocities to those depths. The sonic log velocities at those points are adjusted to fit the check-shot velocities. Between check shots the sonic log provides detailed velocity information. The combination is a calibrated sonic log, like the integrated sonic log, but more correct than could be made from the sonic log alone. Like the simpler one, it is a sonic log at its normal depth scale, but with added marks indicating seismic times at ten-millisecond intervals. The marks provide a relationship between depth and time. The time to a horizon can be read directly from the time scale at the depth of the formation top on the log.

The times to reflections on the section can be calculated from this information. But first an adjustment must be made for the difference between the datum plane of the section and the elevation of the kelly bushing on the rig that drilled the well.

The best correction is made from a reflection that is recognizable as being from a formation that is visible on the sonic log. A limestone, for instance, with a considerably higher velocity than sands and shales overlying it will appear as a large change in velocity on the sonic log and as a strong reflection on the section. The best correction to apply to the time on the log is whatever one will make it match the reflections on the section.

If you can't be sure of the correlation, then you will need to calculate the correction from near-surface data or assumptions. Correct the log to the section, rather than the section to the log. The section is hung from a datum plane that is uniform over an area, while the KB elevation is a measure of the elevation of the ground at that point and the height of the rig floor. Make the correction as described under "Datum Plane" near the end of this chapter.

SYNTHETIC SEISMOGRAM

A synthetic seismogram, that is, a theoretical seismic trace, can be made from one or two of the logs in a well. If it is properly made, it can look enough like the traces on a section to be correlated with them.

Data processors make a synthetic seismogram from a sonic log or, better, from a sonic log and a density log together. A sonic log measures the velocity of sound in the rock. A density log measures the density of the rock. A seismic reflection is a reflection of sound from an interface between two types of rock, differing from each other in sound velocity and in density. The greater the difference, the stronger the reflection. The effect of velocity is the greater, so the sonic log alone is adequate if there is no density log.

The log or logs are used to calculate the reflectivity, the ability to reflect sound, at each change in rock type encountered in the logs. These reflectivities are then used with a seismic wavelet, representing an impulse from the sound source, in calculating what seismic wiggles would result from that wavelet's traveling down to and being reflected from those interfaces. The results of that calculation, plotted as a trace, make up the synthetic seismogram to be correlated with the section.

To make the synthetic convenient for the interpreter, the synthetic trace is displayed at the same vertical time scale as the seismic section. Depths are also indicated where they fall on this time scale. The trace is repeated several times, to look more like a section rather than a lone trace. Several other displays are also shown. The sonic and density logs and a log of the calculated reflectivities are included, also at the vertical time scale of the section. The repeated trace is shown in normal polarity and again in reversed polarity. It may be shown in different frequencies. It is displayed with only the primary reflections. It is shown separately with only calculated multiple reflections and also with both primaries and multiples on the same trace. The whole set may be displayed in several versions, calculated using different wavelets. These versions are used because matching the wavelet to the section is a matter of trial and error, and there may later be a replay of the section that a different wavelet might better match.

In using the synthetic, you first need to mark a geologist's picks of formation tops from the well onto the depth scale of the synthetic. This depth scale is not uniform, as the time scale is the one that was made uniform to fit the section. Then, if the synthetic is minimum phase, check against the sonic log to see how much lag there is from a change in velocity to the point at which the reflection has become prominent enough to pick. This lag will be on the order of 30 to 60 ms (milliseconds). Compare the various forms of the synthetic trace with the seismic section, to see which most nearly matches it. Your decisions will be between the two polarities, between the different frequencies, between the different wavelets, and between the primaries-only and the primaries-with-multiples forms. The multiples-only form will not be one to consider. It is there just to help you recognize the mutliples.

Having selected the best-fitting synthetic trace, lay it on the section and shift it up and down to get the best match. You will already have done this some in selecting between the playouts. Then mark the formation tops on the section. You now have takeoff points to continue picking the section. And don't worry about the different starting points of log and section. Your shifting has already taken care of that.

Using the synthetic seismogram adjusts for any timing problems— but only if the synthetic and the section are both of good quality and look enough alike to correlate correctly. If some of those conditions are not so good, you have to depend in part on the timing of the synthetic to put the synthetic trace in about the right place on the section for correlation.

A good synthetic seismogram, that is, one that looks like the section, makes for a very good identification. When you have correlated a section with a synthetic seismogram from a well that is on the seismic line, and the two look alike, and a wiggle on the synthetic that represents a certain formation top matches one on the section, you can feel pretty sure that **that** reflection was reflected from **that** formation top. Also you can correlate a synthetic with sections farther from the well, and you may obtain good identifications there. Away from the well location, correlation works but datum correction does not.

Identify the reflections on a section by using a synthetic in Exercise 4–3.

VERTICAL SEISMIC PROFILE

There is an even better means of identification. At present the ultimate in reflection identification can be made from a vertical seismic profile, or VSP, which is similar to a short piece of seismic line but is shot with the geophone spread distributed vertically in a borehole. The information recorded, after some rearranging, shows reflections in section form, both outward from the hole and down below it. The reflections can be identified on the VSP from geologists' picks of formation tops in well logs. Then the VSP section can be correlated with a regular seismic section from a line that crosses the well.

VARIOUS CLUES

Other clues to identification are some of the subject matter of seismic stratigraphy. Alternating sands and shales produce many reflections. Massive bodies have few reflections. Alluvial deposits have jumbled reflections. From these relationships and others that may fit the specific situation, you can often do fairly well at identifying at least some reflections on a section.

In all methods we have gone over, from synthetic seismograms to hints at identifications, I have acted more or less as though there was only one well in your area. Of course, the more wells the better. An identification that looks good at one well might be shown to be incorrect in the light of information from other wells. Don't make an identification from one well alone if you have others available. Your identification will be much sounder if it is based on a number of wells.

Now, what about the areas with no wells and no previous shooting? A virgin area, maybe in a country that is just awakening to the possibilities of oil within its borders, presents the most difficult identification problem.

With no wells, it is necessary to find whatever clues there may be. If there are outcrops, it may be possible to trace a seismic reflection from the vicinity of the outcrop. The reflections won't meet the outcrop; we don't get seismic data right up to the surface. But the nature of the outcrop may provide some clues. A difference in rock types may indicate a large velocity difference. Depositional conditions may allow some of the seismic stratigraphy clues to help.

If there aren't even outcrops to work with, as in an offshore area, you are pretty solely dependent on the seismic data alone. Information derived from stacking velocities gives some clues to lithology. The seismic stratigraphic evidence of deposition helps. And carefully worked-out seismic stratigraphy may determine the area's cycles of uplift and depression of the land. The horizons can then be fitted into a known worldwide pattern to determine geological age.

Other clues come from other geophysical techniques, if any have been used in the area. Refraction seismic lines yield more exact velocity data than reflection shooting, although much poorer depth information. Gravity and magnetic data give information primarily about depth to and configuration of granite basement.

OTHER FACTORS

We have taken up a number of ways to identify reflections. There are other factors to be considered.

DATUM PLANE

The sonic logs, synthetic seismograms, other seismic sections, etc., that are used to identify sections sometimes need a datum plane correction to make them fit the section you want to identify. The section you are doing the identifying on has its zero time at its own starting point, a datum plane or, on some offshore sections, an unadjusted start near the surface of the sea. The data from the two, the source of identifying information and the section with horizons to be identified, must somehow be made to start at the same point. This is usually done by adjusting the other information to the same starting point as the section. You adjust that information rather than your section because it is much less work. The identifying information comes from one to a few points of contact, like wells or line intersections, but you are going to handle data all along your line. You may of course be adjusting one line to another.

A seismic section from a line shot on land usually has both the elevation of its datum plane and the velocity used to correct it to that plane noted in the header. If you need to change a datum to a lower one, you will need to subtract the same amount from each of the two parts of the sound's path, down and up. So divide twice the distance you are moving the datum by the datum correction velocity, to give a correction time, and subtract that time from a reflection time on the section. To change a datum to one at a higher elevation, go through the same operation, but add instead of subtracting. You can then match the section with the other data by times, by shifting them so the new reduced or increased time matches the timing of the section that needs identifications. A good way to correct a synthetic seismogram, etc., is to calculate the correction for your section and then apply the correction in reverse to the synthetic or other source of information.

IDENTIFICATION QUALITY

If there are several wells with identification data in the area, assemble the data from each of them and a seismic line through or near each well location. Use all the data you have, but weight it in favor of the best data. If you have VSPs that look enough like the sections to correlate with them, that VSP information should be believed before other information, and so on down the line of diminishing quality of identification. Even though you have some very good means of identifying, it is worth your while to at least check the poorer means you have at other wells. Failure to check something might lead to a surprise later in the project. For instance, suppose you have three wells with synthetics that correlate nicely with the sections, but later in your interpretation you discover a major fault, with all of the synthetics on the same side of the fault. It would have paid you to have looked at, say, a well across the fault that didn't have a synthetic but did have a velocity survey.

You also need to pick some of the horizons from one well to the next, on the most direct routes along seismic lines. This gives a seismic tie between the wells, to show whether your identifications agree with each other.

As neither the seismic tie nor the identifications are certain, checking them against each other is better than using one alone. If they disagree, you can try to work out the problem by deciding which might be changed with the least likelihood of being wrong.

You may, of course, have only one well or none in the area. Then you can't make such a solidly based identification. Your ability to identify will vary greatly from area to area. In each case, use all the information available for that area, which may be anything from good solid identifications down to very little information.

CHAPTER 5

Work an Area

Now we're down to the reality of interpretation. What you have to interpret to find oil is not a line or a loop, but a whole area. In the process of interpreting the area, you may find one or more leads, any of which may develop into prospects, then into drillable prospects, and then into oil or gas fields. So let's get to work on the area.

PICK THE SECTIONS

An area to be interpreted consists of a number of lines. The lines may be all those that were shot in one continuous period of shooting. Or, if that was not the first time the area was shot, it may include the lines that were shot in earlier programs. Or, if the data processing for a recent or current shooting program is underway, more sections may arrive from the processors from time to time.

When you start interpreting, it helps if you clear off your desk. You can use all the room on it. You may even want to put the telephone in a drawer or on the floor or on a windowsill. We will assume you have been given an area to interpret.

WHICH HORIZONS DO YOU PICK?

The choice of horizons to pick is affected by several factors, so it is not a simple matter of picking the reflection from the layer that is expected to produce.

The first step in choosing a horizon might be for you to look over a few sections scattered around the area, to get an idea of what horizons have good reflection quality in all or most of the area. An obvious rule can be stated here:

You can't make a good interpretation from a poor reflection.

From the well data and talks with the geologists, you can select several formations that would be desirable to map, then identify the reflections that represent them. If some of the formations turn out to be at reflections that are good in the area, you can decide to pick and map them. But maybe some of the useful formations do not appear on the section as good reflections. What's best: to pick them anyway, pick something else nearby, ignore that zone?

Either of two things can make a good reflection, a strong contrast in velocity of sound (and rock density) or a series of lesser contrasts that happen to reinforce one another. If there isn't much difference between the velocity of a formation and the other formation lying on it, then there won't be much of a reflection at the interface between the two.

A poor reflection may be picked with effort, but it may also mislead you. The formation may be the one that produces oil in that area, but if its reflection is so poor that you can't pick it reliably, then you aren't accomplishing much in the search for oil by stubbornly trying to pick it. In this situation, a better reflection nearby will probably give you an interpretation that is more representative of the configuration of the formation you wanted to pick than a difficult, and unreliable, struggle to map the formation itself. By looking at the section, you can usually tell which nearby reflections conform with the one you would like to pick. If all the reflected energy between the two, and what you can see of the poor one, are going in the same general direction, then you can be fairly confident that the two layers conform.

Suppose, though, that you have some overpowering reason to interpret the exact formation. One reason might be that one person in your company has the final word on which prospects will be drilled, and that person insists on basing the decision on maps of the producing horizon. Another reason could be that it is difficult to tie wells in the area, and you feel that you must try to tie the very horizon, not something else. Then you will be picking the poor reflection and will be subject to the problems of picking it incorrectly. To guide your interpretation, it might then be sensible to also pick some better nearby reflections above and below your reflection, if they appear to conform to it. They at least provide boundaries for your picking of the poor horizon. If your picks wander astray, you know they can't break through the other reflections. And you can be fairly sure that the intervals between your reflection and the others should remain fairly uniform, at least in proportion. If the reflection is about a third of the way down from the upper horizon to the lower one, then it is likely to remain about a third of the way, even though the distance between the two may become wider. This applies as long as the three formations appear to conform.

We have been choosing horizons in terms of difficulty. Assuming those difficulties are taken care of or that the reflections are all fairly good, let's think about how many horizons should be picked and how those should be chosen.

One of the main factors in the decision is how much time is available for the interpretation. You might feel that interpreting is so important that it should be sort of timeless, taking as much time as necessary. But other tasks, including interpreting other areas, are also important. That timeless approach can rarely be used in oil exploration.

You may be told to pick just one horizon, the important one for the purpose of the moment. Fine, pick just the one. Otherwise, you may be able to select the horizons that will give a good idea of the geology of the area.

Most vital is the pay, the formation that, at other places in the general area, is known to contain oil. Select the reflection from that formation, or a reflection nearby that you think will give you control for the pay formation, that is, be nearly enough parallel to it to substitute for it.

If there are several potentially productive formations, you may want to pick each of them. Or, if they conform well to each other, then one or two may give information that can apply to all of them.

It is usually good to pick a shallow horizon. There is likely to be some shallow horizon that is a very good reflector. An interpretation of it can help you to decide how valid your deeper interpretations are.

The basement is also good to interpret, although often difficult. A basement map can be helpful to geologists in working out the geological history of the area and therefore finding oil prospects.

Summing up, in an area that you want to interpret fairly thoroughly, you will probably pick three horizons—shallow, producing formation, and basement—and perhaps more, to fit the geological situation. But in many areas you may only have time to pick one horizon.

LOOP BY LOOP

Earlier, we went over some details of ties on one loop. When you have an area that includes a number of loops, interpreting is just doing one at a time but with the different loops influencing each other. Select some loop as a starting point, preferably one with good reflection quality. Tie it, by resolving whatever problems make it difficult to tie. Now go to a loop adjoining the one you have already tied, and thus sharing one side with it. Tie it the best you can, changing the first side if that makes the best tie. If the second loop "unties" the first, then work the two out together, arriving at a solution that makes both of them tie. Similarly tie a third loop. Continue with additional loops until all the loops in the area are tied. You will probably have had to make a number of compromises in the process and may not be totally satisfied with them.

Sometimes when you are picking along on a good reflection, you suddenly come to a place where it splits in two. What do you do, take the high road or the low road? Well, the split occurred because the rocks changed. Real geology is causing the problem. Your decision is not one of which path is correct. They're both real. It is one of which you want to interpret. The rock layer you were interpreting may have become thicker, as it reached a different depositional environment. Another layer may have pinched out against it. There may be an unconformity with an uptilted eroded layer ending at your reflection. So you only need to decide which reflection to continue with or perhaps to pick both. Then when you pick other sections in the area, you will need to remain consistent from section to section. In Exercise 5–1, pick a reflection that separates.

When you finish an area, if it is a large one that took a long time to work, you are likely to feel that, with what you have learned about it, you could really interpret it better if you were to immediately start interpreting the area all over again. You probably won't have the time to.

There are things that you can work out loop by loop and other things that must be ironed out considering the area as a whole. Regional faults and major uplifts can be considered best in terms of the area overall. Some interpreters prefer to work out the fault pattern and other regional features before beginning the detailed picking. They then draw these features on the work map, and only after that do they start picking individual loops. This is a good approach, especially in areas that are distinguished by large regional features. In plainer areas, with less tectonics, it may be better to just work a loop at a time and let the regional characteristics develop as you go.

FRAMEWORK

In an area where there are two or more wells and you can identify some reflections at the wells, you can connect the identifications with seismic sections. Those sections then form a framework of picked reflections. These are your most definitely identified picks. They can be on the most direct lines you could get between anchor points at the wells, even though they may zigzag badly, maybe going east, south, and then east again to connect two wells. Those lines constitute a framework to build your interpretation on. It is probably safest, that is, it will probably result in the fewest false starts, if you build up your interpretation of the area with a loop out from one of those lines, then another loop using one of the framework lines as one side, etc. If at some point you decide your interpretation is all wrong—a common occurrence—then you can retreat back to the framework lines, or even repick them, and start out again. It will probably mean less re-doing than if you had started out from the first framework loop and headed immediately away from the framework.

This framework is a powerful tool in seismic interpretation. All the wells in an area should be tied together with such a framework in order to save interpreting time and make interpretation more correct. If there are more than three or four wells scattered about the area, you may find it helpful to connect some of them by more than one route. Then you can tie these framework lines in large loops before undertaking the detailed ties of loops in the overall interpretation.

These framework lines are particularly important lines in working an area, especially a large area, which has many chances to go wrong. So when re-processing for some lines in an area is being considered, some of the greatest value may be derived from re-doing these lines. They may enable you to see some relationship in a different way, where wells are available to confirm or deny ideas.

Similarly, when shooting is being planned in an area, it may help to violate whatever grid pattern of seismic lines is being used, by shooting straight lines direct from well to well. And if a few years of progress in seismic techniques go by after these direct lines have been shot, and then another seismic program is planned, it may be useful to shoot these well-to-well lines again.

FALSE STARTS

In interpretation, false starts are inevitable, if the interpretation is to be a good one. There are several reasons for this.

☐ Every area is different, so the approach used before probably won't work on the new area.

☐ Seismic interpreting is very complicated, involving many problems and many ways to attack them.

☐ The main reason is that you have to get really familiar with an area to work it well. Trying one thing and then another is an excellent way to really know the area.

So it pays to interpret on a basis of "planned false starts." That is, start interpreting the area with the knowledge that you probably won't pick all the sections, then time them all, and then map the times, nice and straightforward. Recognize that instead you will probably start, decide you're on the wrong track, start over, stop again, several times. To prepare for these new starts, don't go straight ahead. For instance, don't color all of a long line before you tie loops. The reflection may look good, but your picks won't necessarily tie when you later complete the loops along the line. And maybe you don't go on with loop ties without putting some data on a map and contouring it to see how things look. You might start picking a good reflection that, on looking the area over quickly, appears to continue to be a good reflection all over the area. But in doing the actual work, you may find a zone of change across the area where that reflection becomes ambiguous and confusing. With the additional knowledge of the area you then have, you may make a better selection of a pick for the whole area.

RECOGNIZE PROBLEMS ON THE SECTIONS

The sections you pick are made up of two kinds of information: the subsurface data you are interpreting and everything else on them, in particular seismic phenomena that tend to obscure or complicate the data. Interpreting the data is in part a matter of recognizing the other information and allowing for it, and in some cases, getting something done about it.

Inherent Problems

There are some problems to be recognized on the sections that are inherent in seismic sections, not put into the data by field techniques or processing. These are some of the factors that make seismic sections not look quite like cutaways of the subsurface. You need to detect the differences and allow for them before you can come to conclusions about the geological meanings on the sections.

MULTIPLES

Multiples are insidious in that they look like reflections and are difficult to remove by field layouts and processing techniques because they **are** reflections. They are reflections that, instead of simply being reflected from a horizon directly up to the geophones, were reflected at least twice. Our ideas about reflections are based on the assumption that the sound goes down and back up just once. But if some of the energy fools around on the way, bouncing off of something again before it shows up at the geophones to be recorded, that extra time doesn't make it very different from an ordinary reflection.

The multiple, having taken longer, appears on a seismic section at a later time, naturally. But we tend to think of things at a later time as being deeper in the ground. It looks like a reflection from a deeper horizon. There are some slight differences in the geometry of its path that distinguish it. It appears farther down the section but did not go through the rock at the depth normally represented by that time. And that rock, being buried under more weight than the shallow rock, is more compacted, so it transmits sound faster than the rock up where the multiple actually did its traveling. This difference in velocities is used to permit stacking to discriminate against the multiple, weakening it somewhat. The normal moveout, that is, the difference in times between paths to geophone groups nearer the shot and those farther from it, is removed before stacking as a standard processing procedure. That difference in times is in part determined by the velocity of the sound. So the correction for the normal moveout (NMO) of primary reflections is not correct for the normal moveout of multiples. Correction of the NMO of the multiples would require a slower velocity than the primaries need. This makes the multiple reflections stack poorly, so they aren't as strong as they would otherwise be.

But that multiple attenuation isn't total, and it varies with reflectivities of formations and with other factors. After the best the processors can do, you may still be confronted with sections that are infested with multiple reflections. You need to recognize them and not pick and map them as primary reflections. It would be embarrassing to map a prospect, see it drilled, and find that the drill encountered granite above the depth at which you thought there was a prospect—embarrassing, expensive, and hard on your credibility.

How, then, do you recognize a multiple for what it is? For one thing, if it appears to dip, it will be steeper than the primary. If the multiple made the primary's trip down and back, but made it twice, then it will have twice the dip. The shallow end is at twice the time, and the deep end is at twice the deeper time, so the apparent dip, in seismic time, is doubled. If the multiple made its extra bounce between two layers closer together, only that part of its dip will be doubled, so it will have more dip than the primary but less than twice as much. That doesn't sound very useful. Deeper geologic formations usually dip more than the shallower ones draped over them, so in that way the multiples look like plain reflections.

Actually, the increase in dip is of some use in recognizing multiples. Measure the times between surface and primary, and between primary and what you think may be a multiple, at several places along the section. If you find them consistently about the same, that may be some confirmation that the alignment is a multiple. Or, if you measure the times between two strong reflectors and see that they are consistently the same as the times between primary and suspected multiple, you may have an interbed multiple that traveled twice between the two strong beds.

Also, multiples may cut across primary reflections that belong at those times. Layers of rock don't cut through each other, so crossing reflections must be explained some other way. One possible explanation is that one of the reflections is a multiple of a shallower primary. Another is that one reflection was reflected from a layer off to the side of the section.

Those clues aren't particularly diagnostic ones, so let's go to some surer evidence. If multiples are suspected, one thing that can be done is to recognize some reflections that are not multiples. Multiples have more dip than their primaries, and they dip in the same direction. So a reflection that dips in a direction contrary to everything above it is obviously not a multiple of anything. Also, a reflection with less dip, or exactly the same dip as the reflections above it, is not a multiple unless the primary is flat. Those clues are limited in application as they indicate only some, not all, of the things that are not multiples; but where they do apply, the clues are quite definite and reliable.

On land, if the terrain is rugged, the plot of shot point elevations on the section can help you to detect multiples. The topography was removed from the section by the data processors, but a multiple that bounced back down from the surface of the ground had that variation in it twice. Only one of the two has been removed, so the multiple, which traveled farther down from the hills than from the valleys, bends down under the hills and up under the valleys, like a mirror image of the topography.

A useful device for detecting multiples by any of the methods above is a horizontally compressed section. It makes dips appear steeper. So it amplifies the various dip effects, making them easier to observe.

REVERBERATIONS

Reverberations are multiples of a special kind that are encountered offshore. The top of the sea is a good reflector, with its large difference in velocity of sound between water and air. The bed of the sea is also a good reflector when it is a hard bottom. Then sound can reflect back and forth between those two. In the worst cases, the whole section, from top to bottom, can be a mass of reverberations. There are special processing systems for reducing them, but the systems do not totally remove them. In interpreting, you can use the shape of the sea bottom, like the elevations on land, to distinguish these multiples. But, instead of being sort of a mirror image, they will dip in the same direction as the sea bed, with the first reverberation having twice the dip, the second thrice the dip, etc.

DIFFRACTIONS

Diffractions are curved alignments on seismic sections that are caused by abrupt changes in reflecting horizons. They look like reflections from curved surfaces, but they come from isolated points, not continuous surfaces. It is important to be able to recognize diffractions so you are not fooled into thinking they are reflections from bent formations. A diffraction is in the form of a hyperbola, curving downward in two directions from its central point, the point of the abrupt change. Its shape is a little like that of an upright (open) umbrella (Illustration 5–1). One change that creates a diffraction is at the point where a formation is broken by a fault. But the reflection from the formation may mask the

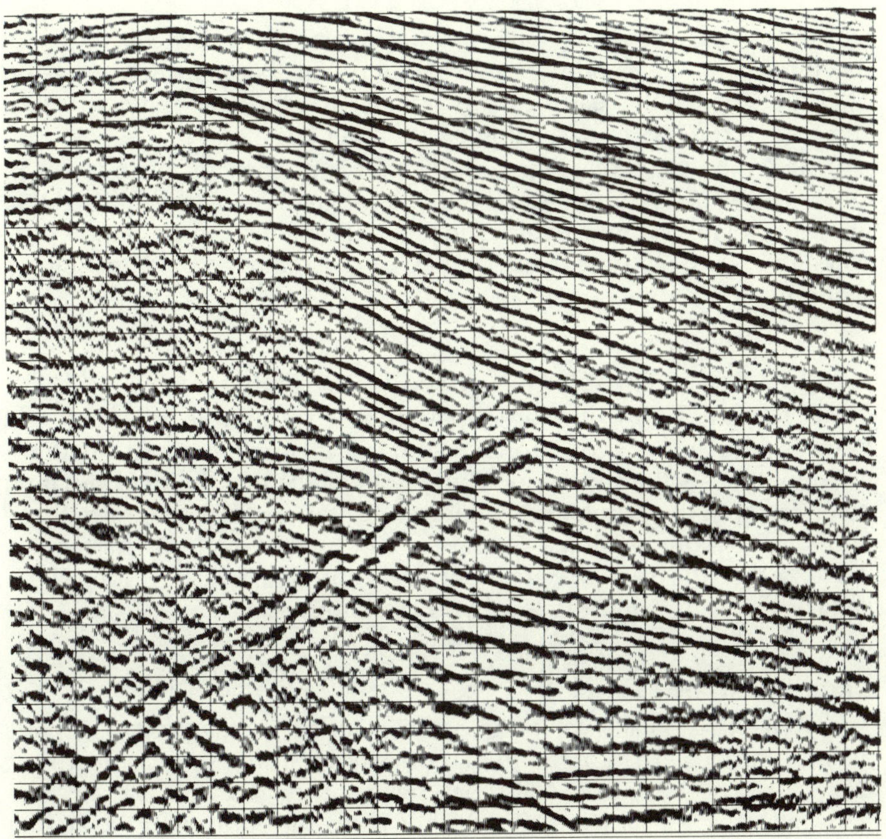

Illustration 5–1 Diffractions
(courtesy Digicon Geophysical Corp.)

part of the diffraction to that side of the fault, so only the part beyond the end of the formation would be visible. These "half-umbrella" diffractions are the most common on seismic sections.

That same situation, a half-diffraction extending from the end of a reflection, can look like an anticline. If the formation is dipping down away from the fault so the fault is at the high end of that segment of the formation, the diffraction looks like dip downward in the other direction. This appearance has tricked many people into thinking there was an anticline in places where there was not. Fortunately, in some of those cases, oil was trapped in the high end of the bed against the fault.

A number of formations may be broken by the same fault, so there may be a slanting line of diffractions or half-diffractions on the section where the fault cuts across it.

An irregular surface—like an erosional surface, irregular reef, or igneous body—has a number of isolated points that constitute the kind of abrupt change that causes a diffraction. An extensive irregular surface can have a roughly horizontal line of diffractions along it.

One key to recognizing a diffraction is its smooth, mathematical-curve appearance. Another clue is the alignment of the curves that indicates a fault or an irregular surface. Diffractions can be handled by recognizing and ignoring them, or they can be removed by migrating the section.

Problems Caused in the Field

Characteristics of the area that are encountered in the field, and the ways the area is shot, produce some effects visible on the sections. Any time the sections are poor, there is a chance that the field situation contributed to the problem. In interpreting, you need to recognize these problems on the sections. Doing so can keep you from mistakenly thinking that the sections are poor for some geological reason, like a fault or diapir down at the level you are interpreting.

GROUND SURFACE

Reflection quality will be poor if the surface of the ground or parts near the surface are of a nature that does not transmit sound effectively. Dry sand, caliche, loose soil, coal, lignite, scoria, gravel are all poor transmitters. They all have air or other gases mixed in with the harder parts, like the sound-insulating material used in office ceilings. The same lithologies, when below the surface, are also poor transmitters. Fortunately, they may not be so bad farther down, as the air spaces become filled with water when they are below the water table. Decaying vegetation, even when under water, as in a swamp, also has gases distributed through it, so it transmits poorly.

These types of surface can change abruptly from place to place. The edge of a swamp, the limit of the outcrop of a formation on the surface, or a change from soft soil to river valley gravel may cause seismic data to undergo a sharp change in quality. Man-made differences, like roads, levees, cultivation, can do the same. Indications of the differences, either natural or artificial, can be obtained from air photos, satellite pictures, culture maps, etc. If such references are available, you should make a point of consulting them. The first reference for such differences, though, is the observer's field report. Observers' reports give statistical information like number of shots taken and in addition have notations about problems and conditions in the field. Notes like "high wind," "hard rain," "cattle on cable," "crossing road," "tanker engine noise," "cable damage" are not only helpful to your interpretation but make fascinating reading.

You can use the information available to you about surface features to avoid misinterpreting them as subsurface features. If you see an anomalous feature that lines up vertically on the section, be suspicious. Look at the ground elevation or water depth indicated at the top of the section. You may have to gather clues over the area. If data is generally poor where the ground elevations are high, there may be looser, drier material on the hills than in the valleys. Or if the poor data is where there are valleys, it may be that the valleys are filled with loose sand or gravel. Sand and gravel transmit sound poorly.

One time, shooting in a river and on into the sea, we noticed that much of the river data was of far poorer quality than the offshore sections. It was natural to think that the shallower water didn't cushion the shots as well as deeper water did. This is generally true. Very deep water gives excellent reflection quality. But in this case, the river data turned out to be poorest in the deepest parts of the river. These deep places were localized spots. Maybe vegetation swept down the river and, becoming waterlogged, collected in these holes and was decaying.

In analyzing the distribution of poor reflections, you may find topographic maps, air or satellite photos, cultivation maps, etc., useful. It may help if you make a map of reflection quality to compare with these maps. If you find some principle that seems to account for the poor records, then that principle can help you in interpreting the bad areas. And when more shooting is being planned for the area, you can determine optimum placing for lines to avoid some bad areas.

GEOPHONE PLANTS

Sound is not transmitted to geophones very well if the geophones are planted poorly. The result on the section is generally poor reflections that cannot be distinguished from the similar effects caused by some surface or near-surface conditions. The geophones may be set loosely on the surface and may not even be upright. Some may even be upside down, sending information of polarity opposite to that of the phones that are right side up. As several geophones make up a group that produces one trace in the field, that trace will be weak if some of the phones are poorly planted, or even contradict others. If the trace is weak enough, the processors may eliminate it from the gather that makes up the final stacked trace. Of course, every trace omitted reduces the degree of stack and therefore the proportion of signal (the wanted information) to noise (the stuff you don't want) on the final trace. The poor plants may be caused by careless or rushed jug hustlers or by field conditions. For instance, on a hard rock surface, what can be done to get a good solid contact? In a howling blizzard, how careful would you be? I suspect I might sometimes be more concerned with numb fingers, feet hurting from the cold, eagerness to get back in a warm truck than in fine detail about just how well a geophone was planted.

LVL

The low velocity layer at the surface, the LVL—mis-called weathering—is a source of problems. If it varies in thickness within short distances, the variation may make the processors' task of correcting the section more difficult and therefore less correct. A special case of variable-depth weathering is glacial drift, the rubble left behind by melting glaciers. It varies abruptly in thickness from place to place, and there may even be two or more layers of it deposited by successive glaciers.

LVL depth is often plotted on sections. You can look at it to see if you can find an explanation for reflection-quality variations. It may be necessary to map it, to make its thickness pattern clearer.

Another problem caused by LVL is in shooting with explosives in shot holes. The LVL, in addition to being slow, is also a poor transmitter of sound. So better records are obtained if the shots are fired below the LVL. If they are shot in it, much of the energy will be wasted in rattling the particles of the LVL. Shot depths are also often plotted on sections, along with LVL depth, making it easy to see which shots were fired in the slow material.

There can probably be a seasonal or weather effect on the LVL. It is dependent on the level of the water table, so a rainy or a dry season, or even a shorter period, in raising or lowering the water table, should also affect the LVL. This may also affect reflection quality.

SEA BED

A similar situation that is encountered offshore is poor reflection quality in areas of thick mud on the sea bed. These muddy areas mostly occur near shore, where the mud collects as it washes off the land. The seismic signal may be overwhelmed by low-frequency energy from the mud. The processors can easily remove the low-frequency sound with filtering, but there may not be much data left on the tape after the low frequencies are removed. A different problem offshore is a hard bottom that reflects energy so strongly it overloads the tape with energy bounced repeatedly off the bottom.

For these offshore problems, as for the near-surface problems on land, the best thing to do in interpreting is to look at the other pertinent information on the section, the water depths, etc. See if there is a pattern that will explain poor data and, if there is, avoid interpreting it as being caused by the deep geology. It may be necessary to make maps of water depth and record quality to see the pattern.

MISSED SHOTS

Data is poor also in places that were not shot, of course. When a land crew is laying out cable and comes to a river, it may have to skip some geophone positions and some shots. But, if the river isn't too wide for it, some of the shots from each side of the river will be recorded by geophones on the other side. There is not a total absence of information under the river, but the stack is not as great as normal for the line. The reflections may deteriorate there. If you are picking and come to a narrow stretch of bad records, you might attribute it to a fault, maybe to help you tie a difficult loop. But before interpreting it as a fault, look at some other information on the section. The elevation plot may show a low place, perhaps even with a notation that there is a river at that place. Also, if there is a display at the top or bottom of the section that shows the degree of stack at each trace, it can give you even more information on why the data is sometimes poor.

Problems from the Processing

Other quirks on the sections may be results of how the data was processed, either flaws in the processing or unavoidable consequences of processing to produce other, desirable, effects.

TIME DIFFERENCES

There are often reflection time differences between the sections from two different projects; so when you correlate intersecting sections from the two, the reflections do not meet. These differences can be caused by several things: the corrections for topography and weathering, the corrections to datum planes, different depths of source and receivers in a marine survey.

When you encounter such a difference in an interpretation, you need to calculate an adjustment to the times of one of the surveys so the reflections from the two surveys will have about the same time at an intersection point. You can't just change a time on a line at each intersection to fit the other line, though. That would be an indiscriminate changing of data. The best thing to do is to determine what computation is causing the difference and calculate a correction for one survey. If that doesn't work, then check a number of intersections scattered over the area and make a uniform adjustment to all the shot points of one seismic program that best fits at an average of the intersections. Then use that adjustment by making a note on the sections, like "subtract 21 ms." When you correlate sections from the two sets, you can offset the timing lines by that amount.

POLARITY DIFFERENCES

The polarity of a section is the direction of the wiggles in response to an upward movement of the ground. Sections can be made with all upward motions shown as peaks or with all of them troughs. This makes for an interpretation problem—two sections of opposite polarities do not match correctly for correlation. If you know the polarities are different, then you can correlate a peak on one section with a trough on the other. But there usually isn't a good way to tell whether the polarities are different or there is a timing difference.

One direction is referred to as normal polarity and the other as reverse polarity. However, there are two opposing opinions as to which direction should be called normal, so a polarity label on a section may be misleading. Also, there are many stages in processing in which the polarity can be changed, so the polarity given on the label may be wrong. The polarity of all the sections in one program of shooting and processing will probably be the same, but from one program to another you can't be sure.

Looking at the sections and trying to compare them, it is very difficult to determine whether the polarities are the same. A half-leg mistie can be accounted for by a difference in polarity or by a time difference between the sections. Being "a leg off" is interpreting jargon for picking incorrectly by the difference between one peak and the next, or one trough and the next. Half a leg is the difference between a peak and the adjoining trough.

A way to try to resolve the polarity problem is to get one of the sections in two forms, one of each polarity (Illustration 5-2). Then correlate each with the other section. You may be able to decide that one polarity matches the other section. When you can't establish the polarity relationship, the only thing that can be done is to treat the difference between the sections as a time difference, correcting the sections of one program to match the sections of the other.

VERTICAL ANOMALIES

A vertical anomaly on a section, that is, any strange-appearing thing that lines up exactly vertically, is suspect. A vertical fault or vertical flank of a salt dome, etc., is possible, but not as likely as a processing problem.

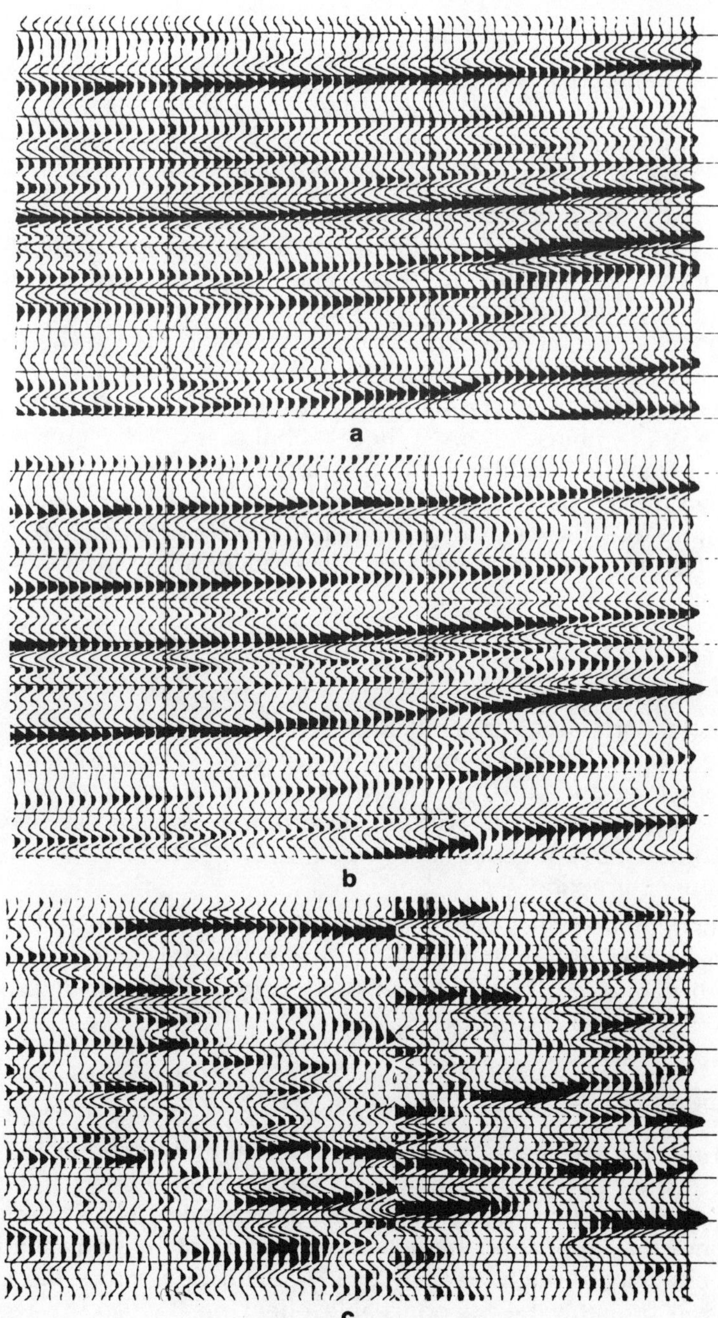

Illustration 5–2 Different polarities:
(a) one polarity; (b) same data, opposite polarity; (c) both polarities
(courtesy Teledyne Exploration)

Look for some change in processing or in shooting, noted on the section. Ask the processors what they think caused the alignment. Look at the phenomenon areally. A truly vertical fault or dike may be a regional feature that you can confirm by finding it on other sections.

There can be some very obvious causes of vertical anomalies on a section. Someone may have wanted two sections joined and so may have taped them together. If the reflections don't join exactly, the break can look like a fault or something. You can of course feel for the tape if the print you are working on was stuck together. But maybe the films were taped together and then the print was made. Or maybe the processors were asked, perhaps against their better judgment, to play the two out as one section. All these techniques make it convenient to interpret from one section to the next; but if you are not aware of what has been done, you may be led astray. If you do discover some joining of sections has created an alignment that is confusing, make a note about it on the section so you don't later have to go through figuring it out again.

STATICS

Static corrections are applied to sections to correct for topography and weathering. These corrections are called static because they are the same all along a trace. They involve just pushing traces up or down to make reflections line up, with no distortion of the traces. They differ from normal moveout correction, which stretches traces and so is called a dynamic correction.

Inadequate statics appear on a section by causing the reflections to have a broken, discontinuous look. About the only thing you can do about this problem is to ask the processors if they can improve the statics (Illustration 5–3).

FILTER EFFECTS

Time variant filters change from one setting to another in discrete steps, so they can produce confusing effects on a section. If a dipping horizon crosses the time at which the change takes place, some slight difference may appear on the reflection. This problem is only mentioned as a possibility, though. Even if your horizon does cross the change, there will probably be no noticeable effect on it. Also, to avoid any effect on an important horizon, people try to arrange to have the filters change at a reflection time where there are no horizons of major interest.

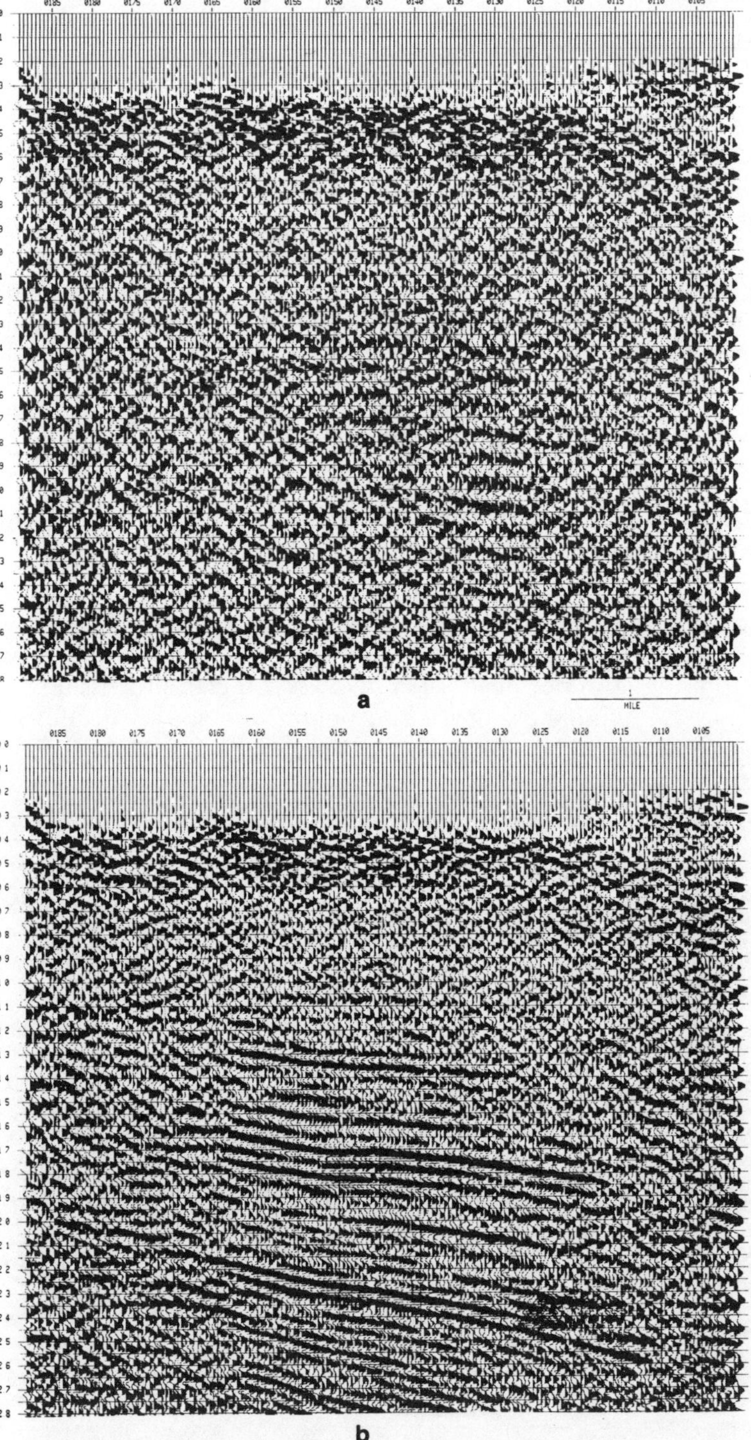

Illustration 5–3 (a) Conventional statics and (b) special statics
(courtesy Teledyne Exploration)

SMILES

Migration greatly improves a section in some ways, but there is a side effect of the migration that you need to recognize as such and not pick as rock layers. The migration removes the down-curving diffractions everywhere that they occur on the section. In so doing, it produces up-curving "smiles" where there is little continuous information. The smiles occur where there are almost no reflections, and the migration reinforces up-curves it finds in random noise. At bottom and sides of the section, where there is no information beyond those edges to be blended with the events within, the smiles are prominent. A person easily becomes familiar with the characteristic appearance of smiles and so can ignore them when picking. Look at the smiles at the bottom of the section in Illustration 5–4.

PROCESSING INVOLVEMENT

Interpreting does not necessarily involve you with processing. The sections may be completely processed before you see them. But just being a geophysicist in an office may cause you to be called on to see

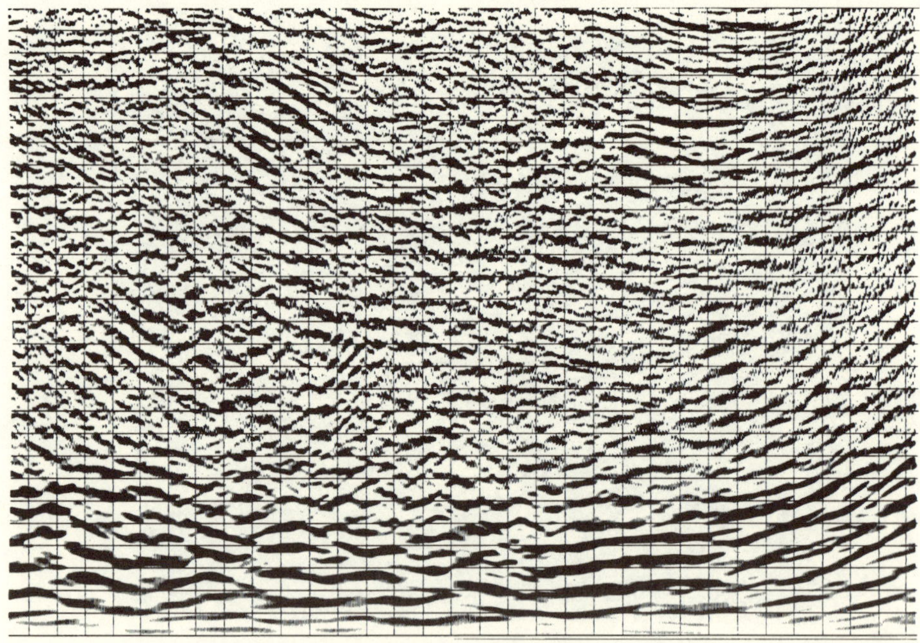

Illustration 5–4 Smiles
(courtesy Digicon Geophysical Corp.)

the processor's representative, to help in choosing parameters for processing data, to offer an opinion on reprocessing sections you already have.

Parameters and Tests

There are a number of parameters that allow choices to be made in processing. They are carefully selected, usually by the combined geophysical brains of both interpreting and processing people. Then, during processing, there are often visits by processors to the interpreters' office, or the other way around, in which the sections are inspected to see how well the selected processing steps are working out.

DECONVOLUTION

Deconvolution is a process to cut down on repetitive wiggles on a trace. This can reduce the vertical spread of a reflection from, say, 100 ms, to maybe 50 or less. The reflections become more distinct from one another, not overlapping so much. With deconvolution of broader reach, multiples, which are also repetitive, can be reduced. So checking the deconvolution on a section is a matter of looking for a crisp distinctness in the individual reflections (Illustration 5–5) or for a reduction of certain multiples. Deconvolution is sometimes applied only before stack and sometimes both before and after. Inspections of comparisons are made to decide if the two stages actually improve the data.

GAIN

The gain setting in the playout of the traces onto a section controls the distance the wiggles swing to the sides from the central position of the trace. If they swing too far, there is so much overlap that there isn't much visible difference between the strong and weak reflections. Thus, it is hard to recognize the character of an individual reflection. If they do not swing far enough, the traces don't overlap enough for the V-A (variable-area) filled-in parts to join and make a continuous band. Also, the weaker reflections may not be detectable at all. I said, "too far" and "not far enough," which are indefinite, but the decision must of course be made by looking at the sections. They should be checked in detail and also in overall appearance, preferably by comparing several sections with different amplitudes (Illustration 5–6).

TIME VARIANT FILTER

The time variant filter used on a section is actually not so much one filter as some, usually two or three, different filters used over different time bands on the same section. They are designed to have the frequencies that are best for bringing out the reflections in the different zones.

INTERPRETING SEISMIC DATA

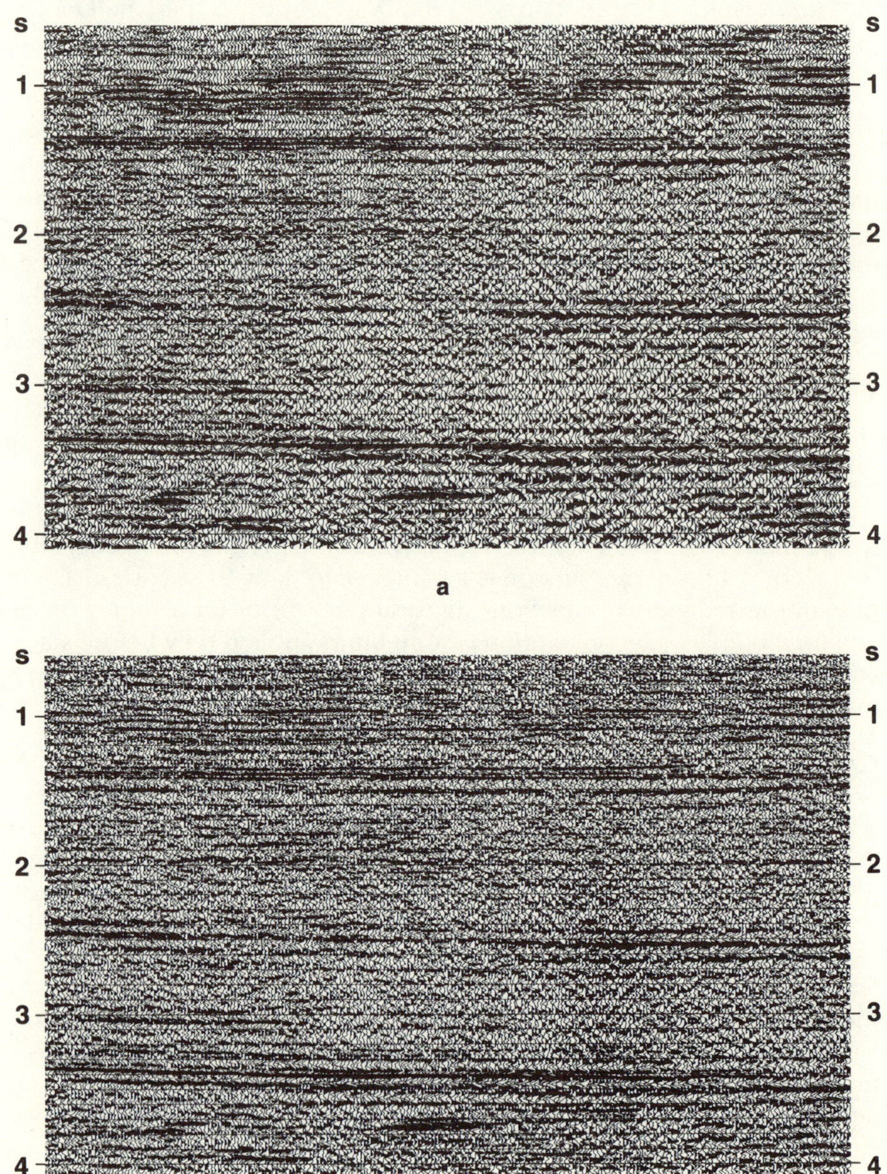

Illustration 5–5 Effect of deconvolution:
(a) no deconvolution and (b) deconvolution
(courtesy Western Geophysical Co.)

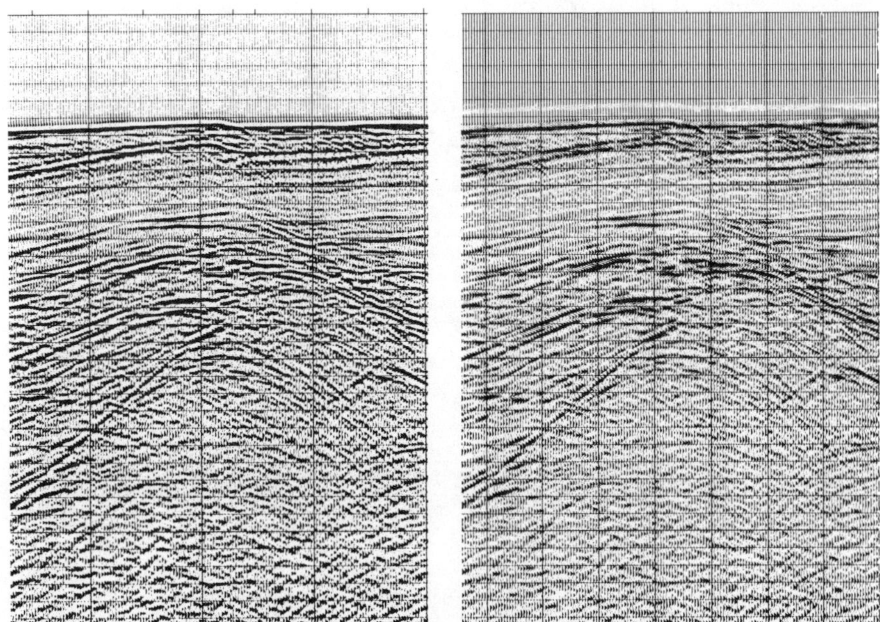

Illustration 5–6 Different amplitudes
(courtesy Petty-Ray Geophysical, Geosource Inc.)

Filters to use are determined from displays of some data at different filters, as in Illustration 5–7. Comparing sections with different filters used in a zone, you will be looking specifically at subtleties of reflections, at whether one filter shows some fine distinction in a reflection better than another.

In general, people like to have all the high-frequency energy that is meaningful. The higher-frequency reflections, being finer wiggles, are closer together and therefore allow the picking of more subtle things, like thinner layers and clues to deposition. But high frequencies that are mostly noise muddle the information on a section rather than adding to it. In comparing filters, look for fine distinctions that appear geologically sensible and therefore probably real. High-frequency noise obscures the real information, but high-frequency reflections are additional information.

VELOCITY ANALYSIS

Before processing, the traces from the farther geophone groups are longer than the traces from near groups. This normal moveout difference must be removed by shortening the longer traces. The correction

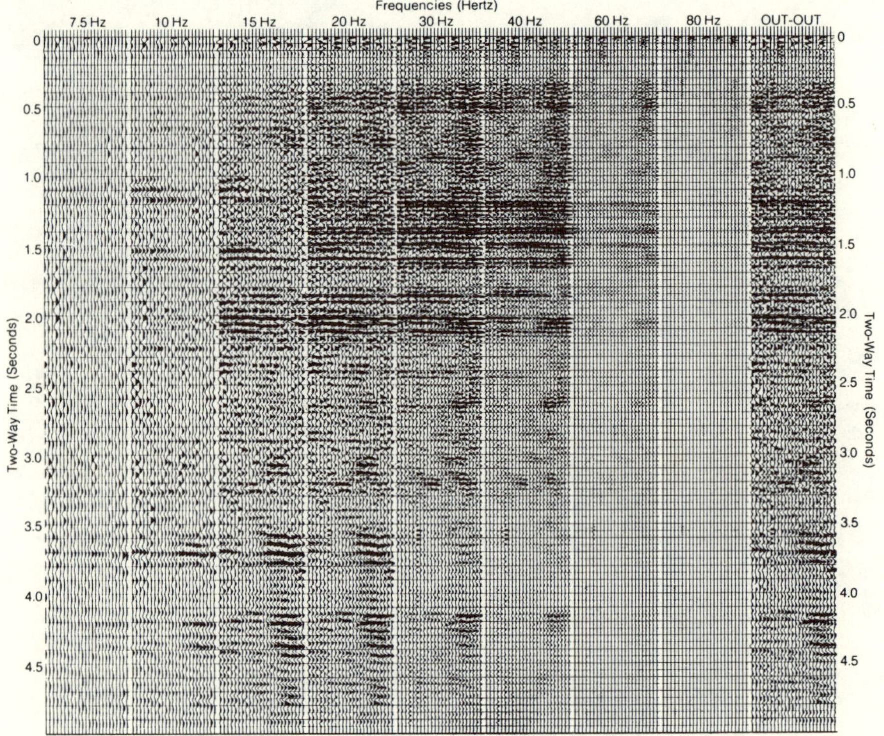

Illustration 5–7 Frequency analysis
(courtesy Teledyne Exploration)

makes all the traces have the same seismic time for one reflection point regardless of the distance between source and receiver. The determination is made by stacking traces at different velocities and seeing which velocity gives a reflection the highest amplitude. Such a determination is a velocity analysis.

The analysis usually takes the form of two displays. One is a set of short bits of section stacked at the different velocities. The other is a graphical display of the amplitudes as contours or wiggles on a plot of velocity against time. The processors may show you their analyses, that is, one or both displays with marks indicating their velocity picks. Picking velocities from either type of display or both together is very subjective. Decisions are influenced by recognizing the difference between primaries and multiples and by smoothing between nearby velocity analyses.

On the velocity vs. time plot, multiples are evident by being directly below their primaries, that is, having the same velocity but a different time. The primaries line up in a curve of velocity generally increasing as time increases, with some irregularities. Velocities picked exactly from the displays would vary wildly from one velocity analysis to the next. This variation is not real and would not be practical to use in stacking data. So the smoothing is used. This means that it is usually not reasonable to criticize a processor's pick of a single velocity analysis alone.

VARIOUS STEPS

Some processing steps—for instance, deconvolution, statics corrections, and stacking— may change a reflection's character or apparent structure. You can check a processing step by comparing versions of the section with and without that step. This may mean that you will need to ask for some alternate playouts. You can also ask the processors about possible harmful effects of a processing step.

DOES IT SERVE THE PURPOSE?

All the above are standard items that may be checked on any seismic program. But you also need to check on how well the processing serves the purpose of the seismic program. There isn't any processing sequence that is best for all sections. Amplitudes, filters, all the parameters, must be set as compromises. The best gain setting for weak reflections isn't the best for strong ones. In order to enhance the deep energy, you may to some extent sacrifice the shallow.

Your lines may have some specific problem, maybe severe multiples or difficult statics that make the reflections poor. Or detail shooting may have been done to answer some specific question, like where the edge of a feature is or whether some reflection pinches out. Look at the processed sections to see whether they answer the main questions, not just whether they look like good sections. Think about the problem and about what you need on the sections to resolve it.

Particularly, discuss the problem with the processors. They can come a lot nearer to solving your problem if they know what it is. Your conversation with them may go something like this:

> "Can you eliminate this effect to bring out that one?"
> "No, but we have a program that will do this; will it help?"
> "Maybe, can you modify it like this?"
> "Sure, if you don't mind this side effect."

That kind of discussion with the processors may be the most useful kind you can have with them. It can get you the best processing—the best for that situation, in that area.

Extra Processing

When an area is shot, a system of processing for the whole area is used—certain deconvolution, filters, etc. It is necessary to keep the processing uniform over the area so lines will tie and the reflections on them will look alike. Then there are some additional types of processing that may be applied to some lines as needed to help solve interpretation problems or help examine prospects or that may even be added to the standard processing package for all the lines.

TRUE AMPLITUDE

"True amplitude" sections are not exactly that. The amplitude of the energy from a shot, thump, pop, etc., varies from large to tremendous at the start, but in a few milliseconds diminishes to extremely small. So a **true** true amplitude section would be a mess of wildly overlapping traces at the start, and below that, it would consist of apparently dead traces, with no visible energy, no wiggles at all that could be seen.

The true amplitude displays that are produced are really relative amplitude sections. They show the energy diminished at the first and enhanced later, so reflections can be seen on all parts of the section. But the amplitudes are correctly comparable from trace to trace and from one time to another. The section is made to show the true amplitudes that the reflecting layers would produce if the sound from the source did not weaken with time (Illustration 5–8).

This type of section, then, is useful for comparing amplitudes to find and measure amplitude anomalies. These are the bright spots and dim spots that sometimes indicate gas in the formations. You may want to have a true amplitude section made when interesting amplitude features appear on the regular sections so you can check them. Or you may want the sections made just to show what amplitude effects might exist on a prospect to make it more drillable or to move the drilling location. In some areas known for productive amplitude anomalies, it may be worthwhile to have all sections played out additionally as true amplitude sections.

MIGRATION

Sound that strikes a reflecting surface perpendicularly is reflected back to its source. If the reflecting surface is horizontal, the sound is reflected from a point directly below the source. But reflection from a **dipping** bed back to the source takes place at a point offset in the updip direction. Similarly, when the sound is recorded some distance from the source, the reflection point from a dipping bed is not midway between the two but is offset in the updip direction.

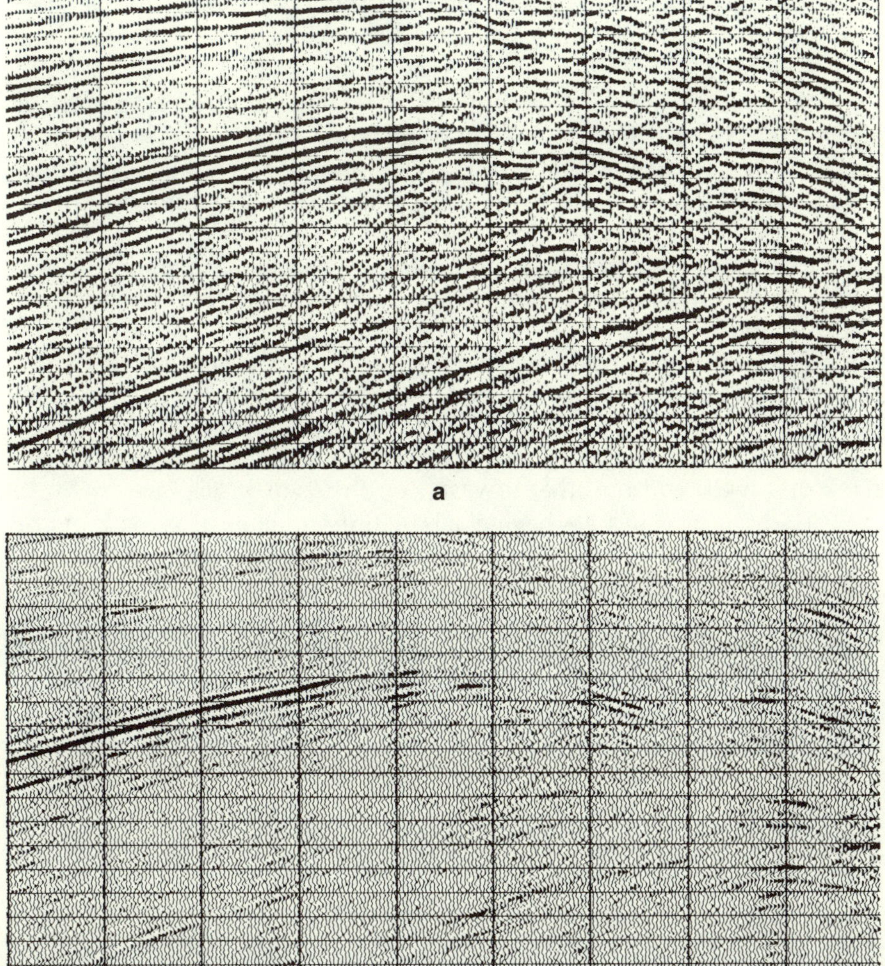

a

b

Illustration 5–8 Effect of true amplitude: (a) normal section and
(b) true amplitude section
(courtesy Petty-Ray Geophysical, Geosource Inc.)

However, seismic traces on a section measure only the travel time.
The traces hang straight down, so it appears on the section that all the
reflections are from points straight down, not offset. Migration is a
processing step to put the correct offsets into the section. It is usually

done in two dimensions, by moving the reflections along the section. This isn't correct if they were reflected from points offset to the side, but it is more correct than leaving the section unmigrated.

Migration removes diffractions, the downward-curving hyperbolic shapes that extend from breaks in reflections. That is, it removes most diffractions but leaves some diffracted energy if the data was not migrated far enough or if the diffracting point was out of the plane of the section.

In shifting reflections to more correct locations, migration makes faults clearer and easier to pick reliably. It also puts them in positions that are more nearly correct. It makes synclines and anticlines, the locally low and high parts of horizons, more nearly their correct sizes, the synclines wider than they had appeared, and the anticlines narrower. It would be bad to drill a well on the basis of an unmigrated section, located on what appeared on the unmigrated section to be the flank of a high, and find that it wasn't on the high at all.

There is a trend to having all sections migrated as part of the normal processing. This is very sensible. Subtleties otherwise unsuspected may be found on the migrated sections. And the migrated sections are a great aid to the interpretation. They help in the detection of faults, pinchouts, reefs, diapirs, etc., and indicate the dimensions of features more correctly. It is usually better for finding oil with minimal waste of money on dry holes to have all lines migrated as a standard part of the processing.

With the shooting done in a straight line and migrated along that line, there isn't any information from the sides of the line. If there is dip to one side, the data should be migrated to the sides also. But the line doesn't provide any information on the dip to the sides.

The solution to this problem is, of course, to add some data at the sides of the line. Parallel lines, cross lines, or some method of giving width to the line are ways to get some data at the sides. A mass of data that blankets an area with closely spaced data points, a 3-D survey, is a more thorough solution to the problem.

With one of these ways of acquiring data to the sides, migration can be performed in three dimensions, so the data can be shifted to the right location. The location is right, that is, if near-surface corrections, reflection picking, and velocity information are correct. But any data, seismic or any other kind, is subject to the accuracy of the raw data and the calculations and corrections applied to it.

After looking closely at a number of sections before and after migration, you get the feeling that you might be able to visualize how a migrated section would look, after just seeing the unmigrated section.

You won't ever become good enough at this to get by without having it migrated, but the ability will help you decide which lines to migrate and what changes migration is likely to make in your sections.

Work Exercise 5–2, to get a feeling for the effects of migration.

MIGRATED LINE TIES

A migrated section is a much more correct representation of the subsurface than an unmigrated section. Even a section that is not in a dip direction has its data mislocation improved somewhat by migration.

But two-dimensionally migrated sections do not tie at line intersections. The unmigrated sections have their data in the wrong places, but they are all wrong in the same way. If a reflection comes from a point to the north of a line intersection, it is misplaced in the same way and by the same amount on a north-south line as it is on an east-west line. So the lines tie at the intersection.

But a line that is migrated in two dimensions has its data in the wrong place in a way that is dependent on line direction. If a line was shot down dip, the migration is about correct. But if an intersecting line was shot exactly along the strike of that reflection, migrating it does not move the reflection at all. Three-D migration permits reflections to tie, but 2-D migration does not.

This is a real problem in interpretation. The migrated sections give a better view of the subsurface. They show the features looking more like the real subsurface features. For some subtle effects, you may have to use migrated sections. But then, when you are ready to go on with interpreting, to tie loops and put data on a map, they let you down and you have to go back to the unmigrated sections.

There are some compromises you can make:

☐ Tie the loops on the unmigrated sections, map the unmigrated data, but also indicate on the map some of the characteristics of features as seen on the migrated sections—exact locations of faults, shapes of reefs, extents of diapirs, etc.

☐ Tie the migrated sections the best you can. Also tie the loops on the unmigrated lines and use them as a guide. Omit some data at intersections if necessary.

☐ A variant of that one is in areas that do not have very steep dip. The migrated sections can tie quite well if the dip isn't steep, so migration doesn't move reflections far. It does remove diffractions, make faults clearer, show sharp synclines correctly. This is the ideal situation for 2-D data.

☐ Map only migrated data—only the dip lines, leaving out the strike lines. Make the correlation between the dip lines agree with ties made on the unmigrated sections.

If there is much dip, any compromise has great flaws in it. But so does any way you handle the situation. You can map unmigrated sections and then migrate the map, but that doesn't use the improved information from the migrated sections. The only real answer to the problem is to shoot 3-D surveys and have the data migrated in 3-D. The trouble with this is that it is **very** expensive. It isn't good economics to use 3-D at today's cost, for exploration work, except in very special situations.

Exploitation is a different matter. When a field has been discovered and large amounts of money are to be spent developing it, then a 3-D seismic survey can be a small part of the cost, one that easily pays for itself in helping to spend the large amounts more sensibly. A 3-D survey can save millions of dollars in otherwise poorly positioned wells or platforms or make it possible to produce tens of millions of dollars' worth of oil that would otherwise be missed.

For an example of the good 2-D situation, tie migrated sections that do not have much dip, in Exercise 5–3.

IS IT MIGRATED?

People usually look at the header of a section to determine what has been done to the data. But you may be asked about it in a meeting, when it isn't convenient to peer at the header. Or the header might be folded under or cut off. It could even have wrong information on it.

What are the distinguishing characteristics of a migrated section? The most obvious, most foolproof sign is the smiles. They form a scalloped-looking bottom edge of the section and sometimes are up within the section. They do not appear at all on an unmigrated section.

First, look for smiles. If there are some, the section is migrated. If not, and if the bottom of the section hasn't been removed or omitted, then it isn't migrated.

Next, look for diffractions. Migration is a process designed to remove diffractions. If there are a lot of diffractions, then the section isn't migrated or was migrated very poorly. A few diffractions may remain on a well-migrated section—diffractions from out of the plane of the section. But, generally, migrated sections are free of diffractions and unmigrated sections have them.

Third, the main reason for migrating the section was to put the reflections in their correct places. An unmigrated section can't show sharp features, expecially narrow synclines. The way the reflections look on a section can help you decide whether the section has been migrated or not.

MODIFY SECTIONS

There once was a saying among geophysicists, "It's only optical." It was used scornfully to describe any technique to make seismic data easier to interpret, that didn't produce any actual new information, but only made it easier to recognize the information that was there. A similar expression now in use is that some effect is cosmetic.

We do need processing that produces additional real information. But, also, much of our processing is directed at effects that are optical. This part of processing also helps find oil. Don't be persuaded to not use some processing for that reason. Your interpreting is optical, too.

Some optical effects that can contribute to finding oil are variable-area darkening of reflection peaks, flattening of sections, and compressing sections horizontally.

Modifications of seismic sections to achieve some of these effects can be performed either by computer or by hand. The computer modifications are more polished, making better sections, but they also take much longer to get done. There are advantages to modifying them by hand. First, if you want to see what might happen if a modification was made, and if you want to see it today, then by hand is the only way it can be done. Second, the one that will concern us now is that, in making the modifications by hand, you learn more about what changes mean than you do in just seeing them already complete. If you gain some experience, say, in flattening sections, you acquire a feel for the effect of flattening on data that will help you in making decisions about having sections flattened by computer.

FLATTEN A SECTION

A seismic section represents a cross section of the earth. If the line is a dip line and the data was processed well and migrated well, it looks like just that—a cutaway view of that part of the earth, except that its vertical dimension is time rather than depth. The geology is visible in its present-day configuration.

But geology is not only a present condition, it is also a process of change through geologic time. It would be useful in interpretation to see the cutaway, but as of some earlier geologic time. This can be done, using some assumptions, with a seismic section.

Look at a section, preferably one with several good, strong, continuous reflections across it. The ups and downs of the reflections are mostly the result of tectonics, movements that took place after the deposition, distorting the beds. If we change the section so as to make a reflection flat, then the section more nearly represents the geological situation at the time that bed was deposited. The part of the section below that bed represents that situation, but the part above it hadn't been deposited by that time and so is not part of the paleogeological picture. We can ignore it or cut it off the section.

Flattening a horizon also makes thickening and thinning more evident. A large change in thickness can be recognized readily on an ordinary section, but small changes in thickness, or the more subtle variations in the large changes, are not apparent on an unflattened section. When a horizon is flattened, differences in thickness, particularly those between it and another horizon, are easier to see. If very slight thickness changes are sought, then it will help if the flattened section is also horizontally compressed. This will exaggerate the changes that take place laterally on the section, so they will show up better.

Flattening a horizon can also make clearer some depositional features that might be difficult to judge without an awareness of which way was up at the time of deposition. Reefs are more easily detected when they can be seen with the horizontal the same way it was when they were formed.

The above reasons to flatten a horizon on a seismic section are geological. They make geological effects more apparent. They also work with other types of data. A section made up of well logs, for instance, is often "hung" on some formation, that is, that formation is made flat to show how other formations appear in relation to it. Conceivably, a photograph of a cliff face could be distorted to make some horizon flat.

But there are other reasons, not geological but geophysical, that sometimes make it worthwhile to flatten seismic sections. These geophysical reasons for flattening have to do with the quality of the seismic sections or with their interpretability.

In some areas the near-surface has velocity irregularities that are very difficult to correct. Glacial drift and permafrost in particular can cause velocities to vary so greatly and so abruptly that correcting for them to make a smooth section is difficult. If all the horizons on a section move up or down at the same shot point so they all look irregular in the same way, this is surely not real geology. If a good shallow horizon is flattened, those irregularities are removed and the section makes better geologic sense. Whatever dip the shallow horizon actually has is thrown away, but shallow horizons are often nearly flat. This error is usually not as great as the irregularity error that has been corrected. So even though it is hung on the shallow horizon, the section can be used like a normal structural section. It is a structural section made with the error of omitting the dip of the shallow horizon, rather than one made with the error of inadequate correction for near-surface irregularities.

The other seismic reason to flatten is to aid interpretation. If record quality is generally not very good in an area or the beds are highly folded, so horizons are difficult to pick with much confidence, then it may help to select the best reflection and flatten it. With it flat, some other reflection, now more nearly flat, may be easier to pick. If need be, that one could then be flattened to help in picking another or others. And so on.

The flattening can be done by data processing. Select a shallow, reliable, continuous reflection and ask the processors to flatten it. A replay of the section will have that reflection at zero time and all the reflections below it shown in their relationships to it.

You can also flatten a horizon by hand. Get an extra print of the section. Pick the horizon to be flattened, marking it with a bright color. Tape the section down to a surface that you can cut on. Cut the section vertically into narrow strips, about as narrow as you can handle—two or three traces per strip works well. Have a piece of paper nearby with a zero line ruled on it, and with enough vertical lines to help you lay the strips vertically. Tape the strips, one at a time, onto the paper, with the flattening horizon on the zero line. Make a copy of the new section on the office copier, so it's all one piece.

Then, if you wish to study further, get another copy of the original section and flatten a deeper horizon. You can make a series of sections illustrating tectonic and depositional development in the area.

I repeat—do it. You will be surprised at how much you learn from such a hands-on use of the data. After we started flattening by hand, it always seemed that getting the work done by processors was too much trouble. We could decide to flatten a section, look at it an hour or so later, decide to change the way the horizon was picked, and do it again, all in one day. That way, with the problem we were investigating still fresh in our minds, we could really make use of the information we obtained from the flattening. In Exercise 5–4, work with a hand-flattened section.

If you don't need the whole section flattened, but just shifting your picked horizons will be good enough, there is a quicker way to flatten without having to cut up a section. Pick several horizons on a section. Get a sheet of no-print grid paper or draw a horizontal zero time line and some vertical lines on transparent material. Lay it over the section, with the zero line over the horizon to be flattened at one end of the section. Vertically below that point, mark the other horizons. At a nearby point, shift the section up or down so the zero line is at the flattening horizon at that point. Mark the horizons below. And so on across the section. Then connect the marks for each horizon. Be careful not to connect two points that are on different horizons. It's easier to avoid this problem if you have just marked a few horizons and those continue all across the section.

COMPRESS A SECTION

Another way of modifying a section to see things in the geology that aren't apparent in the conventional section is to horizontally compress the section, leaving the vertical dimension unchanged. This too can be done either by processors or by hand but is easier and better done by the processors. With it too, though, you can get the feel of the process by doing a few sections by hand.

To compress a section, it isn't necessary for a horizon to be picked first. The processors have several alternatives, with one method working better for one processor and a different one for another. The processor can either space the traces closer together, combine some traces, omit some, or a combination of those methods.

A normal section is made to be easy to pick. It has a long enough horizontal scale to allow individual wiggles of traces to be visible. Its horizontal scale is also sufficient to make it no great problem if an interpreter doesn't happen to be exactly vertical in dropping down from a shot point marked at the top of the section to a horizon to be picked. There are some aspects of the section that show up more clearly if the horizontal scale is highly compressed relative to the vertical scale. A compression of about six to one is a dramatic enough difference to make the section appear quite different and therefore to permit a quite different look at the data. Illustration 5–9 is a compressed section that

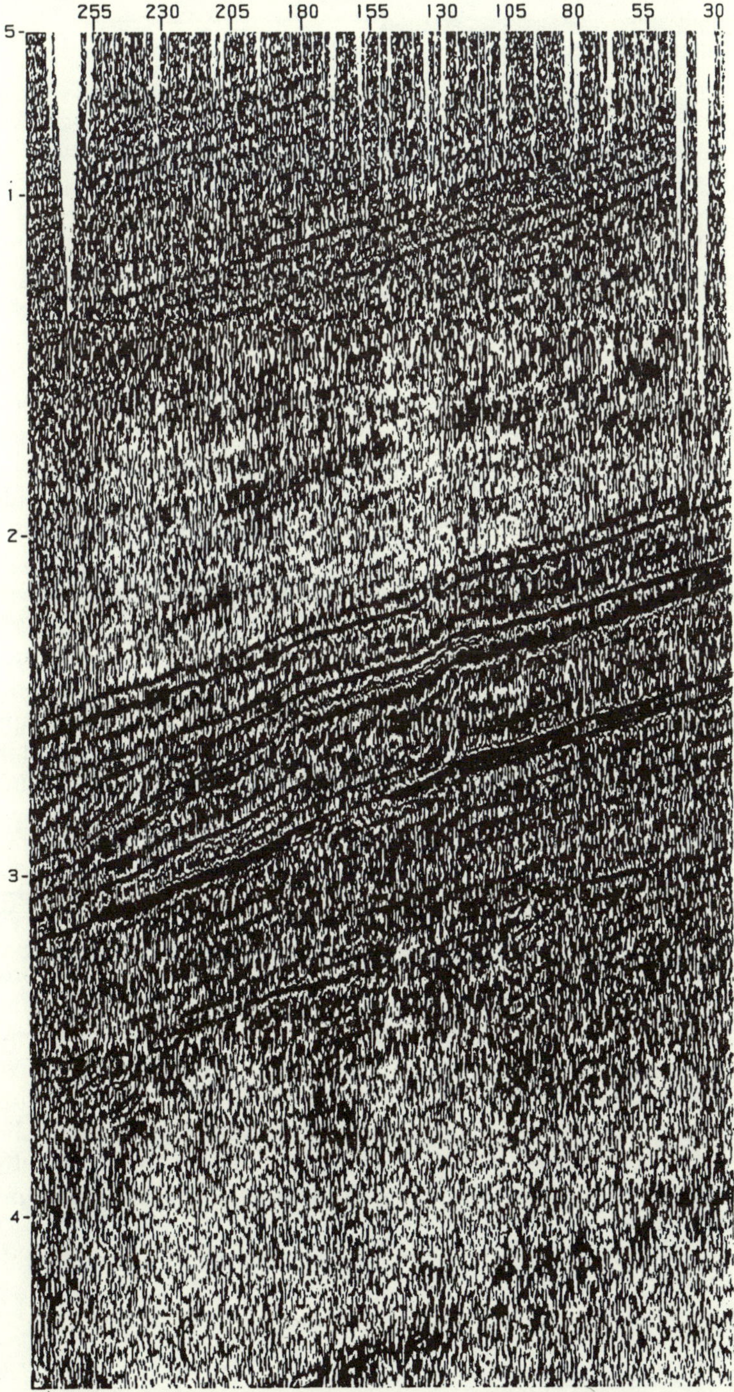

Illustration 5–9 Compressed section of amplitude display
(from color original courtesy Seiscom Delta United)

displays the amplitude of reflections. The illustration is a black and white copy of a color section. The area is one of gentle relief.

Subtle relief, which might not be noticed on the normal scale section, becomes sharper. A wide low-relief anticline becomes a smaller, more abrupt bump. Faults become more vertical and therefore easier to see (on this kind of section, vertical anomalies aren't necessarily suspect). Reefs may show their characteristic shapes more clearly. Regional features are easily recognized on a long line that has been compressed. It is often worthwhile to place a standing order with the processors to compress every line and present the compressed section at one end of the normal section.

The compressed section has defects of its own, but the idea is not to substitute it for the normal section but to provide it in addition, as an alternate view of the data. You can't read reflection times from the compressed sections very well. For one thing, a slight deviation from the vertical in finding a point on the horizon that is below the shot point would make your pick not apply to the shot point you intended. And some features that are clear on the normal section are subtle on the compressed one. An abrupt dip in a reflection might be easily picked on a standard section but on a compressed section might be mistaken for a fault or picked incorrectly by one cycle. Both have flaws. The two together make a great combination. Try the combination by working Exercise 5–5.

Now for the hand compression. Like flattening, compressing of picked horizons can be done on an overlay without cutting a section. Use no-print paper or make your own form. Mark shot points on it, closer together than on the section. Lay the zero line on the zero time line of the section and keep it there. Mark the horizons at one end of the section, then shift horizontally to match positions of the next shot point. There is even more risk of connecting different horizons than on a flattened section, so it is best to have only a few horizons, preferably picked on the entire length of the section.

We have covered some of the main points in dealing with one of the two kinds of tools of the interpreter, seismic sections. We are now ready to go on to the other, seismic maps.

Put It on a Map

Up to this point we have been considering how to pick reflections on sections, but we have not given much thought to what will be done with those colored marks on the sections. There is some geological understanding to be gained by just looking at the sections with the picks marked on them. Anomalies that might become prospects can be seen on them. But the sections are in more or less flat planes, and you need to consider all three dimensions of the subsurface and to have the information in a simpler, more diagrammatic form.

TIME THE PICKS

The most-used way of presenting seismic data in three dimensions is by reading reflection times of the picks at intervals along the sections, putting those times on a map, and drawing contours on the map to make the ups and downs of the horizon easy to see. Reading the times is called timing the reflection, or timing the pick, or, in currently popular pseudocomputerese, digitizing.

You don't need to time the reflection at every trace. That would require a great deal of effort for the information it contributed. For convenience, certain shot points—every tenth, or twentieth, or some such interval—are usually numbered on the section and on the map. Interpreters customarily read times at just those points on the section and then plot those times on the map.

That is convenient, but there is no magic in the labeled shot points. You could time just half as frequently, at every second labeled point, if that gave you enough data for the size of feature you were interested in. Or you might time all the labeled points and the midpoints between them, if you felt the need for closer control. Or you could even use those shot point numbers just as a counting system, and time reflections only at the points of change of dip or other interesting points, like faults or edges of diapirs. A combination that is often used is timing the reflections at the numbered shot points and also at any significant other points: locally highest and lowest points, exact locations of faults, etc.

To time a reflection accurately, you need to have a distinctive part of the reflection to pick. The most recognizable part is the bottom of a trough. It is a point of change in direction of the trace, usually a fairly sharp change. This makes it a definite, easily recognized point, easier to distinguish than a peak because it is not obscured by the solid color of the variable area filling in the peaks. You can pick a trough on one trace quite precisely, but the points bob about a bit from trace to trace, so a sort of average of several traces is a more stable thing to pick. Sometimes you may want to pick a peak as representing more closely the formation top you are interested in. In that case the deflection points, that is, the tips of the peaks, are somewhat concealed by the filled-in peaks on the section. You may have to time just an estimate of where they are, guided by the edges of the dark band and other nearby clues.

There are varying opinions about the amount of smoothing to be used in picking and timing. Some people make the time approximately that of the trough on the one trace at the shot point being picked, taking into account two or three nearby traces. Some make their picks with a straightedge as a means of smoothing many traces together, then time the line drawn along the straightedge. This, though, may smooth so much that small features of possible economic interest are lost. Probably the first choice, smoothing a few traces by eye, is about optimum. The choice depends on the size of feature you are looking for and the quality of the data. The more erratic the reflection, the more smoothing is necessary. The better the data, the more each trace can be relied on individually. However you choose, you may want to make your timing easier by making a dot at the exact point to be timed, before you actually time; or, as most people do, you may take an easier and faster way by making these decisions while you are reading the times.

You can read the reflection times directly from the timing lines, interpolating between the closest ones, usually the ten-millisecond lines. This reading is accurate, even if the paper section has stretched or been otherwise distorted. The closest timing lines are only about a millimeter apart, and there won't be an appreciable distortion in that distance. The reading is accurate, that is, if you don't lose your place among all the lines on the section. The thin fine-division timing lines all look alike, and timing lines are covered in places by the dark areas. There is a considerable risk of misreading a timing line. You can look along one of the heavier lines, at the larger divisions, to the shot point under consideration and then count finer lines to the pick. Where these finer lines are hidden by the dark parts, you may have to estimate the count by vertical distance.

Now do Exercise 6–1, picking and timing a reflection and jotting the times at the top of the section.

A lot of counting and keeping track, wasn't it? And on this section the timing lines stand out well. On some sections the distinction between light and heavy lines may not be so clear, and on some the light lines are omitted.

If you are going to use distance between the heavier timing lines to determine where a point is between them, then you can measure those distances rather than estimate. For that, you need a scale that matches the timing lines on the section. Seismic sections come in several different standard scales, and occasionally there may be a section that has been enlarged or reduced to a nonstandard scale for some special purpose. A quick way to get a scale that fits your section is to use the timing scale at the end of a section—by folding, or cutting off, or making a xerographic copy.

Now work Exercise 6–2, timing some reflections by folding the section so you can use the scale that is at the edge of the section.

Scales used in the United States and often by U.S. companies in other countries tend to be one of two types. Older sections may be derived from standard speeds of tape drives—3.75 inches per second, the speed of a home tape recorder, or twice that, 7.5 inches per second. This is a clumsy system that results in sections that are best read by using a scale made from a section. The other system uses simpler inch-based scales, 10, 5, or 2.5 inches per second. This type of scale can be read with an Engineer's Scale, a manufactured scale that is divided into tenths of inches and other decimal divisions. In most other countries, sections are made to metric scales, usually ten or five centimeters per second. They are much handier—just grab a metric scale and read times.

There is a bit of technique to be learned even about using a scale to read times. The scale you read from and the section may not exactly match. A section can stretch or shrink. A paper scale can, too. A xerographic copier can be out of adjustment. A scale cut from one section may not quite fit another section. When you use a scale and, as is natural, put the zero point of the scale on the zero time of the section, you're not quite ready to read times. You need to look at a heavy timing line near the reflection you are reading, shift the scale so it matches at the timing line, and then read the time.

Exercise 6-3 consists of a section and a scale to use with it. The scale doesn't exactly match the timing lines of the section. Pick the indicated reflections and use the scale to time them.

Even being aware of this problem you might think you could read without shifting, as the error would be uniform. That isn't safe, though. The distortion of the section can vary along the length of the section. Another section may have a slightly different distortion. Even if those differences don't exist, reading incorrect times can lead to difficulty later on, after you have forgotten just how you had read the times or what scale you had used on them. To avoid these problems, match the scale at the top, at zero time; then shift it to match at a point near the reflection. Another method is somewhere between using a scale and reading directly from timing lines. Use a short scale, maybe only 100 milliseconds long, and measure from a nearby heavy timing line.

In the exercises, you've been writing the times you picked at the tops of the sections. This is convenient enough when you're picking only one horizon. But in interpreting there will be many such numbers, for different horizons, re-picks of the same horizon, etc. You need a good way to save all these numbers for use in mapping. One way to preserve the times is to put them on the map as you go. It's best to do so with pencil. You'll make a number of mistakes that you will catch immediately and want to erase, and you will occasionally change your mind about both picks and times. This putting them directly on the map is slow and inconvenient, as you have to go back and forth between section and map, finding your place again each time you look from one to the other.

Another way is to tabulate the times on a separate sheet of paper, preferably with shot point numbers already on it. This, too, involves some going back and forth and losing your place. At least the sheet can be smaller than the map, and maybe you can find room on your desk for both the section and the list.

Either of these methods is made easier if someone else helps you, writing the times on the map or the list as you call them out. Then neither of you has to lose the place. If either one answers the phone, though, or has a visitor, you are both interrupted. A variant of this is to call the times out, with shot point numbers, to a tape recorder. Then you can listen to the tape and write the times on map or list. You'll have to keep stopping and starting the recorder, though, if you read them into the tape too fast for yourself or have many interruptions.

Or you can write the times at the top of the section, as you have been doing, near the shot point number. If you are timing several reflections, you can list them one above another in a regular order. If one of the horizons can't be picked at a shot point, you should indicate the missing time with a dash or something, to avoid later confusion as to which horizon is missing.

Perhaps the most natural and convenient place to write the times is on the section, at or near the reflection. Some geophysicists do this. It has the disadvantage that the numbers are hard to read against the background of traces and timing lines.

While you are timing the sections, you should also be noting any other data that will be necessary on the maps. If the reflection is faulted, you need to mark at least the position of the fault break on that horizon in terms of shot points. You can include more details about the fault if you feel that the data is good enough to have meaning. You can indicate the width of the break in that reflection; the throw, in seismic time; or, instead of the throw, the reflection times of the horizon at the two edges of the fault. If there is a feature—a salt dome, shale diapir, igneous intrusion, reef—that interrupts horizons, you may find it useful to indicate the limits of that feature.

In Exercise 6–4, list the times and the pertinent information about the faults.

In interpreting an area, after the times are listed in some form, they are ready to be put on a map.

MAPPING

Although contouring is the main interpreting part of mapping, you will, in dealing with the maps, necessarily become involved in some physical aspects of the maps themselves. These are all things that can be handled by draftsmen. But if you don't involve yourself in them, then you are forcing the draftsman to make guesses as to what would serve your purposes best and making yourself a helpless victim of those guesses. You and often you alone know what scale will display your data best, what information you need on the title block or an extra legend, how much information needs to be shown on the map.

BASE MAP

A base map is a map with geographical information on it that more specialized information can be added to. A base map may have latitude and longitude, X and Y coordinates, townships and ranges, whatever basic grid system may be used to fix other positions on. It may also have coastlines, rivers, towns, etc. Then, when seismic surveying is done, the shot point locations obtained from the surveying can be plotted on the base map in their relationship to the grid system already on it.

This map then becomes a base map for further additions of specialized information, seismic data. It is a shot point location map, a shot point map, or, to a seismic interpreter, just a base map, as it is the starting map for interpreting.

This seismic base map is usually drawn by the draftsmen from shot point locations provided by the seismic crew's surveyor. The information may be in the form of a map, a list of coordinates, or both. Sometimes, especially in offshore work, a computer-drawn map is provided.

This map could be used without re-drawing, but often shot point numbers fall on top of other numbers or shot point circles, so the map must be re-drafted, or at least edited by draftsmen, erasing and repositioning things that overlap.

The seismic base map is made on a transparent material, usually film. Then diazo copies, in the form of either paper prints or other transparencies, can be made from it.

TITLE BLOCK

Any map needs a label so you and other people will know what it is. The label should be in a standard location on the map so time doesn't have to be wasted hunting for it, for instance when the map is rolled up. There are some bits of basic information that need to be included in every such label, like the scale and who did the work on the map and when, because sometimes those once obvious things aren't obvious any longer, and it's necessary to know them. These requirements are customarily satisfied by the use of a formal title block.

A title block will usually have the following information given somewhere on it:

- ☐ The word "map" to distinguish it from sections, etc.,
- ☐ Company name,
- ☐ Prospect or area,
- ☐ Geographical description (country, basin, sea, etc.),
- ☐ Map scale.

It may also have:

- ☐ Type of map projection used to represent part of the spherical earth on a flat map,
- ☐ Draftsman who drew it,
- ☐ Date drafted,
- ☐ Dates revised.

Maps with data and contours will also have:

- ☐ Horizon mapped,
- ☐ Interpreter,
- ☐ Date completed,
- ☐ Contour interval.

And may also have:

☐   Special notations about the data.

A map made for an oil company by a contractor will have:

☐   Client company name,
☐   Contractor company name.

The form of the title block is a matter of choice. A small or new company, or a contractor, may choose to make it look nice, with all lines centered, horizon and area in large letters, company name a little smaller, and other information in small lettering. It may have a minimum of standard information but have extra notes about special conditions or problems.

Another form of title block is designed like a questionnaire, to make it easy for the person responsible to not overlook any important information. This is often the way of large, or old, organizations. The form is likely to call for more information than the simpler title block— things that had been found necessary in specific situations in the past. But some of them may not apply to a particular map. The completeness is often self-defeating, as people tend to fill in fewer and fewer of the blanks, to the extent of omitting some of the ones that are needed on any map.

A title block is customarily placed in the lower right corner of the map. This location is an industry standard, so you can pick up a map provided by another company and find the title without hunting for it, just as well as on your own company's map.

The main reason a title block is drafted, rather than having blanks filled in by handwriting, is to make it easy to read, that is, with clear lettering and with the main items large enough to be read from a distance. A secondary reason for the drafting is to have the map look professional, so it will give a good initial impression of your prospect or whatever is on it.

The title block usually has the map scale spelled out, like 1:50,000. This is convenient, but not adequate. Maps are often reproduced at different scales. If the 1:50,000 map is shot down to half scale, 1:100,000, there is that 1:50,000 label, lying to everyone. So the map must have a bar scale, a scale shown graphically, with kilometers or whatever unit is handy for the scale, measured on it. Then, when the map is reduced, the numerical label may lie, but the bar scale is there telling the truth. And, if the map was reduced—not to a particular scale, but to some size requirement, say page size for a report—then, no matter how unstandard the new scale, the bar scale is correct. A person can lay a piece of paper against the bar scale, mark the units on the paper, and use it to measure distances on the map; or use a divider to transfer the measurements. Or maybe the map was not reduced, but a much-used print of it shrank or stretched as paper will do. Then, too, the bar scale tells the truth. The bar scale, if it is large enough to be conveniently useful, is too large to go inside a neat title block, so it is usually placed somewhere near it.

SCALE AND SIZE

The scale of a map is in part a matter of availability. You tend to use the scale of map that is provided by the contractor or that is already in the files. Sometimes, though, there is not a map ready-made, or you may need one tailored to your work on a prospect.

The scale should of course be large enough for you to easily put data on your work map, and for it to be easily read on a final drafted map. And it should not be so large that your data points are so far apart that contours appear to have more detailed significance than they do. In general, this means that the shot points that will have data should be at least a centimeter apart and not more than about four centimeters.

The size of a map is another practical concern of an interpreter. There is a limited amount of room on a desk or in a person's reach. For efficient work, your map must fit on this space so that it does not hang far over the sides of the desk or have to be folded or rolled up in your lap. You should be able to reach almost all parts of the map, to see all parts fairly well, and with a little folding to be able to do careful lettering and contouring on all parts of it. The smallest size that there is much point in working with may be about letter size.

PLOTTING SHOT POINTS AND WELLS

There are many shots or pops in common-depth-point shooting. Stacking makes a seismic section continuous, so the location along the line of one pop is about as good a place to put data as any other pop. There is no particular need to plot each pop on the map. The meaning of "shot point" becomes blurred. In a particular seismic survey, it may be decided that every so many pops will be one that is called a shot point. Even those shot points may be too close together to warrant plotting all of them on the map. It may be convenient to plot every tenth, or twentieth, or so, shot point. Maybe all will be drafted as circles, or some as circles and those between as dots.

The shot points are only convenient places to plot data. Any place along a section would do, but it helps you know where you are if points at even intervals along the section are marked and those same intervals marked on the map. Then a pick from the section can be readily plotted at its proper location on the map. This sounds obvious, but there are seismic surveys for which, say, every tenth shot point, starting with one, is marked on the sections—1, 11, 21,. . . and every tenth shot point, but starting with zero, plotted on the map—0, 10, 20,. . . .

Shot point numbers are indicated on the map, not necessarily at all plotted points, but frequently enough so people can count forward or back to locate a certain point.

Sizes of circles, dots, lettering are dictated by practical considerations. Shots and shot point numbers should be drafted large enough to show up well on the map, but also small enough to leave plenty of room for data to be added to the map without cluttering it.

In addition to shot point numbers, each line has a line number, which is indicated on the map at one or both ends of the line and any other place that may be necessary to make things clear. It is customary for each line to have its own set of shot point numbers, independent of the other lines, that may be the same as numbers on other lines. Each line might start with Shot Point 1, for instance.

If there are wells in the area, they need to be shown on the map. It wouldn't be good to find a prospect and then learn there was already a dry hole on it that had been left off the map. Wells are indicated by conventional symbols that indicate their status: oil well, dry hole, gas well, etc.

Shot points and whatever wells there may be must be plotted accurately and in their correct relationships to each other. They can be plotted accurately by careful use of the survey data used in positioning both shot points and wells. The relationship to each other is better known if the seismic surveyors took readings on existing well locations and if the wells that were located on a basis of seismic data were staked by surveying from shot point locations, rather than from grid coordinates.

Wells are in a different situation from shot points. There are usually a great many shot points on a seismic map, and very few wells. A typical map may have some thousands of shot points and perhaps three wells. The wells are especially important to the seismic interpretation and to seeing the seismic data in the context of other information on the map. It is necessary for the wells to be easily and quickly found on the map, so the well symbols should be highly visible.

If the wells are represented by little circles about the size of the circles used for shot points, any of several bad things may happen:

☐ Someone you are showing the map to may ask where some well is, and you may have trouble finding it;

☐ Someone looking at your map may not realize there are wells in the area and so may draw a wrong conclusion;

☐ Worst of all, you may fail to notice a dry hole on your prospect.

Indicate wells by big circles, using the symbols for their status, whether dry, oil, gas, etc. Put their correct names on the map. Naturally, if there are many wells in the area, for instance if there is a developed field, then there is proportionately less need for prominence for each well. You may not even need to show some of the wells in the interior of the field. Use good judgment. Most seismic maps are in areas with not more than a few wildcats and no fields. Those areas need to have the wells displayed boldly.

It is also good to make the well symbols more prominent by adding color to them. Don't depend on that alone, though. At any time someone may make a new print of the map and not bother to add the color. For that reason, the well should stand out even without the color.

ORIENTATION OF NUMBERS

In putting numbers on a map, a person's natural inclination is to put them all right side up so they are easy to read without turning the map. The title block and some other information are best oriented that way.

At one time, when seismic maps usually didn't have as many seismic lines as they do now, it was customary to plot the shot point numbers right side up. Some maps are still made that way. Try Exercise 6–5, locating shot points on such a map.

A seismic map may have many lines shot in various directions, so they cross at various angles on the map. It may happen that Shot Point 150 of Line 23 is near Shot Point 150 of Line 201. With the shot point numbers all lettered the same way up, it can be difficult to decide which is which. From necessity, most of the industry has adopted a practice of putting the shot point numbers at right angles to the line. Then it is easier to tell which label belongs to which line.

This argument is a good one for shot point numbers and is even better for seismic data, as many more data points are put on a map than shot point numbers. Data may be plotted at every tenth shot point, and the shot point number given at, perhaps, every fiftieth. It is even more important to put the data at right angles to the seismic line than the shot point numbers. This applies to you in writing numbers on a work map and to the draftsman in plotting them on a final map. In both cases confusion and misunderstanding are reduced.

CONNECTING SHOT POINTS

Seismic maps tend to get cluttered. As more lines are shot and added to the map, they have to be fitted in with the ones already there. Lines may cross at various angles, some nearly parallel. Even with the shot point numbers perpendicular to the lines, it begins to require time and effort every time you want to find a shot point on the map.

A map may look like the one in Exercise 6–6, before you work the exercise. Don't work it yet; just look at it for now.

Confusing, isn't it? And wait till the draftsman crowds in data at all the shot points. Also, the line numbers would have to be found way over at the edges of the map.

There isn't anything there that can be removed without reducing the information and causing greater confusion, but clutter can be reduced by adding more marks to the map, by connecting all the shot points of a line. Then the lines stand out, and a shot point where several lines cross isn't so ambiguous.

Some people are reluctant to add all those lines to a map; instead, they just put short bits of line at the circles that are near intersections. These bits of line point toward the next points on the lines. It seems like it should work, but it never seems quite good enough. When all the points are connected in lines, the eye sees some lines, rather than a mass of circles, so the map really does seem less cluttered.

When the circles are being connected, there is considerable judgment necessary in deciding which circle belongs to which line. A straightedge is needed to see which circles line up, and a scale of normal shot point spacing for that line is needed to see which circles are at about the right distances from others of that line. If a draftsman assigned to connect the circles does not clearly understand how seismic lines are shot, that is, in fairly straight lines and with fairly uniform spacing, then you may find that it saves you time in the long run to help with the decisions.

In Exercise 6–6, connect all the shot points in each line.

Helps reduce the confusion, doesn't it?

WORK MAP

A work map is any map that you do your interpreting work on, usually one the draftsmen will then use as a source of both data and coutours to put on a final map. It is usually a print of the shot point base map. This is convenient, as it has the same background information that will be on the final map. The work map must have all the shot points, so you will have places to put the data. The map should also have all the wells in the area, so you will know they exist and so you can refer to them as your interpretation proceeds. The map is like work clothes, in that it does not need to look dressy but must be sturdy and serviceable for practical use. You will be doing a lot of marking on it—erasing, adding new marks, leaning on it, folding it different ways to work conveniently on different parts of it. It may get pretty old and worn while you work up your interpretation. Doesn't matter; there will in most cases be a prettied-up map made from it later.

Your data on the work map should be in some more permanent form than the contours, so you can erase contours freely without fear of erasing data. For this purpose you will probably put the data on with ink. You don't need to spend extra time and effort making the work map pretty, so the inking can be done by using a ball-point or fiber pen or whatever is convenient to you. But just a minute. You will probably change your mind about data also. Maybe it is better to put it on with pencil and then go over it with the pen before contouring. Or you may want to put the data on in pencil and then make it more permanent by copying the map with the office copier and contouring the copy.

WHAT DATA IS DATA?

Now, what data should be put on a map before contouring? Seismic reflection times are the main thing people think of as seismic data. But there are other things in the same category, everything obtained from the sections that will influence the interpretation. In addition to reflection times, this includes fault crossings, throws of faults, formation zero edges, amplitude anomalies, edges of reefs, sometimes reflection quality or other seismic attributes. Whatever information from the sections is to be used in mapping should be considered as being data just as real as reflection times and should similarly be put on the work map in permanent form before contouring.

In particular, fault crossings are an essential part of the data for contouring. When you are putting reflection times, also make some kind of mark on the map to show just exactly where a fault was found on the section. The mark should not attempt to indicate the direction on the map of the fault trace. You don't know that yet. That is the kind of interpretation of the data you will make in contouring. Either have the mark cross the seismic line at a right angle or choose a type of mark that doesn't indicate direction, like a circle or square. If you can tell the horizontal distance on the section in which the formation is missing (for a normal fault) or overlapped (for a reverse fault), then that too can be good data to show on the map. If you can read the throw on the section in seismic time, then you may want to add that to the map. It can be indicated by writing the number of milliseconds of throw or by showing the reflection times of the horizon on the two sides of the fault.

Another useful type of data is the reflection times at critical places, whether they happen to be at plotted shot points or not. If something important to locating prospects, like a narrow reef indication, happens to be between those shot points, show it on the map, too. It may help you to determine the shape of a larger reef complex, or whatever. Don't be a slave to the shot point interval that happens to be marked on the section.

This brings up an alternative way of putting data on a map. You could, if you preferred, not put reflection times at plotted shot points, but just at locally high and low points, and at a few intermediate points if the highs and lows were far apart. This might save considerable timing and plotting of data and would indicate just where the highest and lowest points were. Contouring data of this sort requires that you recognize that there aren't any high or low points between the ones you have plotted. A disadvantage of the method is that it isn't convenient to list the data by shot point for manipulation like subtracting to make time interval maps. Also, special care is required to determine a point on the section and that same point on the map. For this it helps if the map is at the same horizontal scale as the section. Then the top edge of the section can be laid along the seismic line on the map, and the points transferred straight from section to map.

PEOPLE MAKE ERRORS

I'll let you in on something that the general public doesn't know and can't know, no matter how often they are told. And you probably won't believe me if you haven't been through the experience yourself. You may believe it intellectually and believe it about other people, but deep down, you don't believe it about yourself. Here it is.

When you work with a lot of data, you will make a number of mistakes.

Any time new people have joined an office or an operation in which a large amount of data is handled and in which there is a way of checking that is routinely used, they have been pretty confident that they worked carefully and therefore didn't have mistakes in their work. Then, in the checking, the first error is found. They are a little embarrassed but consider it a fluke. When the final count is in, they refuse to believe they were so careless. They want a recount, have trouble withholding anger, feel they were tricked. They were tricked, but tricked by the volume of data, not by the other people in the office. The new people feel they don't make errors. The old hands also make mistakes but, because of experience, considerably fewer than the new guys. And, also because of experience, the old hands are aware that they make mistakes.

Any time you work with a quantity of data, and especially when you are new to an operation, try to work out some ways to check yourself. Develop an idea of what is reasonable. Maybe an unreasonable-appearing answer is correct after all, but the odds are against it. For your own protection from yourself, check it two or three different ways if possible before letting it stay. Also, if someone questions it later, you will already know the checks you have applied and so can consider the problem intelligently with the other person. And if you were correct, you will have evidence that you were.

My way, after having learned about myself, is to work out some kind of cross-check that will point out the errors. I know they are in there. Just putting the two factors together, me, and a bunch of data, means that there will be a percentage of errors, some of them pretty troublesome. It's not that I know this because I'm so smart, and it's not that I make the errors because I'm so dumb. It's that I'm a person, and there are a lot of opportunities to blunder.

The best way to check is some system that checks several steps at a time, rather than one by one. As an example, you may write a report by hand and then type a copy of it. You could proofread it by comparing the handwritten and typed copies word by word. This finds errors in the typing but not in the original. Also, it is an inefficient check because it is boring, and so it is hard to keep your attention on each word. A broader, easier, and better check is to read the typed copy for sense, to see if it says what you had in mind. This checks both steps at once. It might miss a typing error that happened to make as good sense as the original, but that kind of error doesn't do much harm.

Similarly, reflection times are often read from a section onto a list, then copied from the list onto a map. If you check the map against the list, you aren't checking mistakes in the list. A more thorough check is made by checking the map against the section.

CONTOUR THE DATA

Seismic interpretation consists mainly of picking reflections and contouring data on maps. Now we've got the picked and timed data on the maps. Contouring will make it usable for finding and defining prospects for drilling.

Contouring is a simple process.

Consider how the surface of the ground might be contoured, automatically and correctly, by physical forces. Let's take an area behind a dam, so we can imagine the water level being raised and lowered to order. Close the dam to prevent flooding downstream and let the area behind it fill with water. Then let enough water out to lower the water level ten meters, and let it stand awhile. Then lower it another ten meters. Some floating grass and twigs will have left a high-water mark at the first level. Continue lowering the water ten meters at a time. Then take a picture of the area from the air, directly above. In the photograph, the series of high-water marks will appear as a pattern of nearly parallel lines around the hills and valleys. The high-water marks in the photo are contours of the topography at a ten-meter contour interval.

A similar effect is seen in looking down from an airplane on an area of contour plowing or of terraced rice fields. They, too, are marks at certain levels, but they aren't necessarily at equal intervals.

Now suppose you don't have the air photo or the high-water marks but just elevations obtained from surveying at a number of points scattered irregularly over the hills and valleys. You can get the same kind of contour map by plotting the points on a map, writing the surveyed elevations at the points, and drawing the pattern that the high-water marks would have made if they had existed.

You have to do some guessing, since you don't have the elevations all along a contour, like the high-water mark provided. You just have whatever points were surveyed. If two points have elevations of 43 meters and 38 meters, then the 40-meter contour has to run somewhere between the two. Obviously. When the water level was at 40 meters, all the 43-meter points were on dry ground and the 38-meter points were under water.

That's a convenient way to start contouring the data, just by drawing a line between the points that are above 40 meters and those that are below it. If some surveyed point was at exactly 40 meters, you'd be in luck. This is right on the high-water mark, so the 40-meter contour could be drawn right through it. The same procedure would give you contours at 30 meters and all the other 10-m intervals. There's more to doing good contouring, to using judgment about the shape the high-water marks would have, but you'd have contoured the map.

The reason for the judgment is that your scattering of points might not be adequate. There may be areas the surveyor couldn't get to or didn't have time for. You'd have to make guesses there.

Now let's go through some very basic exercises in contouring. Let's take a part of the map of the land behind the dam, with some surveyed points but no high-water marks to help us. Figure 6–7 in the workbook has a few points with their surveyed elevations. Now do Exercise 6–7, drawing the 40-m contour.

You have now contoured this small part of the map. For contours at every ten meters, this part only needs one contour, as it doesn't have any points above the 40s or below the 30s.

Now the next figure, in Exercise 6–8, is a larger area of the map. The points you already contoured are in the southeast corner. Contour this larger area.

You probably contoured the first map with a crooked line angling across the map. The second one likely has a contour that goes completely around the high point of data in the 50s. This is a closed contour. It comes around and meets itself.

The next figure, in Exercise 6–9, is a still larger area of the map. Contour it at a 10-m contour interval.

You probably got a pattern that shows a hill with high points.

Let's go back to the water marks and think a bit about what kinds of patterns they form.

- [ ] On a perfectly round hill, the marks will be circles, the smallest at the top of the hill, the next lower one around it, and larger and larger circles to the bottom of the hill. From the air it looks like a target.

- [ ] An elongated hill will have a pattern of more or less oval shapes.

- [ ] A long hill with two round peaks will have sort of oval contours at the bottom, two sets of circle contours at the tops of the peaks, and dumbbell-shaped contours in between.

This wasn't in the maps you contoured, but a river valley will have contours generally paralleling the river, getting farther apart where the river widens and closer together where it narrows.

Now consider the shapes of the contours: the circles, ovals, dumbbells. Notice that, in general, a bulge on one side is accompanied by a bulge on the other side. Or, to put it another way, one side of a hill is more or less a mirror image of the other. This leads to a generalization:

Contours at the same level tend to be mirror images of each other.

But then look at the 40-m and 30-m contours. One goes around the hill and so does the other, leading to another generalization:

Contours at different levels tend to parallel each other.

These two principles are not often stated, but the shapes are so familiar that they are part of people's inherent way of looking at contours. In particular, contours that do not follow these principles look strange and are misleading.

All of that was about contouring the surface of the earth. In giving the elevations of the hills, we talked about meters above sea level. On seismic sections, though, reflection time is measured downward from a datum plane, or if it is converted to depth, then the depth is below some datum, often sea level. The difference here is not so much in how contouring is done as in how it is thought about. In the land behind the dam, 50 m was a high elevation. In seismic data, the larger numbers represent deeper points. Sure, you know that and understand it, but it still takes a mental effort to recognize without having to think about it, that the small numbers are the high points, the prospects, the things we are looking for.

You need to become so accustomed to thinking of lesser times or depths as representing high points that your automatic reaction to a small number on a map is that it is good. To help the process along, we'll do exercises on subsurface data. One of your duties to yourself is to consciously think of the small numbers, especially those surrounded by larger numbers, as high and as possibly being prospects.

As a first such bit of familiarization, Exercise 6–10 has another set of data, this one in terms of seismic reflection time, in seconds of two-way time, carried out to three decimal places. In Exercise 6–10, contour the map and, after you finish it, think about where you might want to drill a well.

You probably decided that somewhere on the high might be a good drilling location, but that there wasn't enough data to allow you to be confident of an exact location. In real seismic exploration you might recommend more shooting. You can't always get all the data that might be helpful, for any of several reasons. There is a limit to the amount of money that should be spent on a prospect before it is tested. That limit is more or less at the point where it becomes evident that it should be drilled. Refinements to make sure of the exactly best spot will turn out to be a waste if the prospect, when drilled, is dry. Another reason you can't get all the necessary data is that there may not be time or budget to allow extra shooting; then later, there might be a rush to select a location so an expensive rig doesn't have to stand by idle.

In Exercise 6–11 is another map to contour. It has a complication that hasn't been in the other maps. See how it looks when you contour it.

You probably got some kind of alignment across the map, where contours made a jog, bending sharply and then bending back to the direction they had been going. This is how a fault looks on literal contouring of the data. But a fault is an actual breaking of the rock layers. It is contoured better by drawing a line where you think the break is and then by drawing the contours to that line. In some ideal situations, the contours look as they would if the map had been cut apart, shifted, and taped together in the new position. It isn't always that neat, but you can draw contours right up to the fault and let them stop at it. Now, in the exercise, erase the part of the contours with the offset, draw in the fault, and run the contours to it, without the bend they had before. If the structures on the two sides look like the same feature, but offset, it may help to draw them that way.

What you have just done is one of the ways to find faults in seismic data. Sometimes they can't be seen on the sections, so contouring is the only clue there is. Usually, though, there are some indications on the sections, which can be supplemented by contouring evidence.

Try another situation; work Exercise 6–12.

You may have found another fault, this time at the crest of a structure, so the high appears cut apart and shifted.

Here is another map to be contoured, in Exercise 6–13. In this case, the data are mostly scattered like they were on the other maps. But in one part of this map, there is a dense assemblage of data, as might be found on a prospect that has been shot several times in a continuing effort to determine whether it is worth drilling. Work Exercise 6–13.

How does that look? You may have an unreal-looking set of contours. There are two philosophies you could have used in contouring it. If you tried to position the contour lines at exactly proportionate locations between data points, the dense part of the map has intricate contours. Flat layers of rock, later distorted, do not take shapes like that, so the contouring doesn't make sense geologically. On the other hand, if you just let a contour go anywhere between data points above and below it, the mapping looks fairly smooth and geologically more sensible. Why should that be?

ONE WAY TO CONTOUR

Let's think a bit about the difference between seismic and geological data. You may have noticed that, in the land-behind-the-dam situation, I said to draw contours between, say, the data in the 30s and the 40s, without saying where in between to put the contour. We could very carefully draw the 40-m contour exactly half the distance from a 38 to a 42, or two-fifths of the way from a 38 to a 43. The reason we didn't at the very first was for simplicity. But there is a special reason to not be so exact with seismic data. Surveying of elevations of the surface of the ground can be done very precisely, to a few centimeters in ideal situations. Depths in a well can be measured to about the same precision. But there are several vagaries in seismic data that make them considerably less precise. Offshore or in a shot hole on land, the depth of the pop or shot may not be known accurately. Near-surface datum plane corrections are made to the sections by applying only an estimate of the velocity of sound. The picking by interpreters is made by reading the position of a trough that may vary from trace to trace, so the picking is an average of several traces. All these inaccuracies together make for several milliseconds of variability in the data. The reflection times at one shot point may be several milliseconds too high, and at the next shot point too low. So treating the data on the map as very exact numbers is pointless. As someone said, "It's like measuring a hog with a rope and a micrometer." The points have a few milliseconds of error, so it is more realistic to contour them by smoothing than to contour them as precise data.

There are different ways of smoothing inexact data. One way is to contour, with smoothness of contouring given priority over the data. Contouring is made smooth by ignoring the last digit or so of some of the data. This produces smooth, geologically sensible maps but is a very subjective process. It is difficult to check, to separate the deliberate ignoring of data from the errors in contouring. And it may be that some of the irregularities in the data that are ignored are indicating prospects.

Another smoothing technique is used in picking the sections. Instead of picking the reflections where they appear on the sections, smooth lines are drawn, often with a straightedge, through the reflections on the sections. At intersections the ties are made exact by placing the straight lines on the two sections so they match exactly. This, like the contour smoothing, is a subjective method that is difficult to check and may miss some useful subtleties in the data. Both work well for some interpreters in some areas, but the methods are difficult to learn to do well and difficult for another interpreter to confirm.

A method I came up with to resolve the problem is to do some smoothing by using the data as exact numbers when deciding which numbers to put a contour between, but ignoring the data completely when deciding how far from a number to put the contour. I put a 40-meter contour, not half way between 38- and 42-meter data, but anywhere between them that happened to make the contouring smooth. This smoothing allows the map to be geologically reasonable, without making it so subjective.

When I had been using this technique for some time, a trainee geophysicist brought up a good point. He asked me why I was giving more respect to round numbers than to the other data. The numbers that were the same as the contour levels, like 40, I was honoring exactly, with the 40 contour going exactly through the 40 points, although a contour could go anywhere past a 41, as long as it went between the 40 and 50. He was right; I was handling some points in a special way. Also, if there were several points scattered about with the same value, say several 40s, the contour would have to snake around unrealistically to go through all of them.

The solution was simple. Think of the 40 contour as a 39.9 contour. Then it doesn't go through any points. Also, contouring and checking are easier. Just make the contour separate all the numbers in the 30s from those in the 40s. The smoothing is also better. That clump of 40s is all on the same side of the contour. To check your contouring, run your finger along between two contours, for instance the 40 and 50 contours, saying to yourself, "40s, 40s, 40s." Anything found between them that isn't in the 40s is contoured wrong.

The process is mechanical enough and predictable enough for a person to easily check the contouring for errors. In checking, there are no subjective factors to consider. A point is either between the appropriate two contours or it isn't. This method also makes the really troublesome points stand out, indicated by strange crooks in the contours, so they can be checked and corrected if they have plotting or calculating errors or so they can be reinterpreted if that is necessary.

Try re-contouring the map in Exercise 6–13 that way—with contours not touching any shot points.

How was that? Easier, probably. When interpreting for a company, you will contour whatever way you like best, but it is worth your while to see how this method works.

CONTOUR AND ERASE

Now let's take up another factor in contouring. A well-contoured map looks smooth and realistic. How does it get that way? Well, I hear that some people are so skilled at visualizing and drawing that, just from looking at the numbers, they can imagine how the contoured map should look. Then they draw the contours—drawing each contour just once, in the right place and with the right shape. I haven't met one of those people yet, or, anyway, haven't seen one in action. Among ordinary mortals like me, contouring can't be done by just drawing. It requires drawing and erasing, over and over.

Some people contour by a modification of the visualizing method. They look at the numbers and try to determine the shapes the numbers represent, then draw those shapes. They find some high (small) numbers and contour that area as an anticline, then notice a smoothly sloping monocline and contour it, etc. They don't try to be exactly right the first time and are willing to do a lot of erasing and re-contouring.

A third way, the one I prefer, is to first draw lines that separate the data—crude lines, even straight in places or shaky. Then, when the data are all separated into strips between contours, the map has been contoured, not well, but completely, and in one sense correctly. Then look at the map, find an especially bad-looking contour, erase it, and re-draw it, still between the same numbers, but trying to make it more like the contours on the two sides of it. That necessarily makes it smoother.

When that contour has been re-formed, again pick a bad-looking one and make it like the ones flanking it. And so on. At first you are making a contour look more like its neighbors, while the neighbors are still crude. That doesn't put it in its final shape. One beside it may later be drawn more like it, then it may need to be re-drawn. You may get back to the same contour several times. After some work, without needing any particular artistic ability, you have the map smoothly contoured and your lap full of eraser crumbs. An extra benefit of this method is that, each time, you are working on what is then one of the worst contours, so you can stop at any degree of perfection you wish or have time for. At any time in the contour reworking, the map is complete and has reached some stage of smoothness. You start with it quickly complete and spend time on improvement. In working with deadlines or bosses who may pop in to see the map, this is better than having a map look very good in one part and not yet contoured at all in another.

A trick to make the contouring look even better is to contour on a closer interval than you want, get that contouring smooth and finished, and then erase the extra contours.

Now work Exercise 6–14, contouring and erasing.

Your map's pretty smooth, isn't it?

INTERPRETATION PROBLEMS

A strange-looking contour is one thing that can point out interpretation problems. One of the most frequently encountered is a mis-tie of two intersecting lines. Contour the map in Exercise 6–15, letting the contours bend sharply if the data calls for it.

Well! At several intersections there were contouring problems. Maybe that's trying to tell us something about the data. Notice that one particular line is involved in a problem every place it crosses another line, and the other lines have problems of this sort only where they cross that line. Sort of points the finger of suspicion, doesn't it? Look at what it would take to eliminate the problems. If the data on that line could be changed to match the other sections, at each of the tie points, then of course it would tie them all exactly. But if we allow ourselves to change data whenever we want, we might as well draw whatever contours we want and not bother with data. The mis-ties are indicating a problem of some sort with that line, though, and we need to correct it. The first step is to look at the section carefully and see if:

- ☐ The wrong horizon was picked,

- ☐ The times were read incorrectly,

- ☐ The polarity is different from the other sections,

- ☐ Some near-surface correction is different,

- ☐ Anything else is wrong or different.

We don't have the section in this exercise. Assume that there wasn't anything that we could find on the section to help with this problem. But there is still a problem. The one line, in this case, is shallower at all the ties than the tie lines. To start with, we absolutely know the subsurface of the earth isn't like that. A layer of rock isn't at one depth at some point when we shoot one line and at another depth at that same point when we shoot another—unless there's been an enormous earthquake between those times. So what is the most reasonable thing to do? Push that line down, probably, even though we don't know why it is too shallow. We must assume that there is some difference in the way it was shot or processed but that we don't know what the difference is. We couldn't justify making it tie at each intersection individually, so we'll push it down by the same amount all along the line, an average amount that will bring all its ties into tolerable, i.e., smoothly contourable, differences. This isn't completely correct, but it is more correct than leaving the data as it was and more useful than throwing out the data from that line. Erase the contours in the vicinity of the line, change the data on the line by a uniform amount, and re-contour.

Better, isn't it? You may have decided that the difference was about 20 ms. That's probably what I'd have decided. It's close. What I actually did in making up the exercise was to make the map with all the data correct and then shift the data on that line by 19 ms. The adjusting technique isn't precise enough to be correct to the millisecond very often, though.

Now, in Exercise 6–16 there is another map to contour. Again, there is one line that is a problem. See what you can do to resolve it.

In that case, raising or lowering the data for the whole line didn't make things enough better. On this map, there is a location problem. Try shifting the location of the troublesome line. If you move it ten shot points along the line toward the southwest, the data fit the other lines nicely. A survey problem could cause this. Such problems occur more with old data, especially old offshore data, than with new. And shifting the location of a line is a pretty extreme thing to do as part of the interpretation. You are not likely to be that correct in placing it. Again, something is wrong. The ties are impossible as the line is. And in this case, with the direction of movement being along the line, the problem could be one of labeling. The processor's draftsman could have labeled the shot points wrong on the section, starting at a wrong point and continuing to count from there. Or the draftsman who plotted the line on the map could have made a similar error, happening to get one point wrong and counting from it. When you decide the error is a matter of location, then, if you have the time and data, you can investigate drafting, processing, surveying, to determine what the error is. The need to shift the line in interpretation is a clue, but it is far better, if possible, to have the error corrected at the stage where it was made.

Okay, Exercise 6–17 is another map to contour, another one with a line causing a problem. What can you do about this one?

Couldn't solve it? I couldn't either. This is a case of a line that is so different from the other lines that the only thing that can be done, right now, in interpretation, is to throw it out. There are the usual investigations that can be made, if you have the time and opportunity. Maybe the line number is wrong on the map or the section, and the line you are looking at belongs in a far part of the area or in another area. Maybe the shot points were plotted backward on section or map. Maybe the shooting or processing somehow went wild. But when you are interpreting, you often have a deadline. Sometimes the information on shooting and processing isn't available to you. You investigate if practical, throw it out if not.

In these exercises with problem lines, I have assumed that the data somehow got picked and put on a map without the problem being discovered. That is unlikely, as you would normally run into it while trying to tie lines. Even so, as a way to solve the problem found in the ties, you might have to resort to putting the points on a map and contouring in order to determine just what type of problem you had. Also, this is a way for you to encounter this type of problem without having to pick all the sections of several areas. What the contouring in these exercises points out can be applied to mis-ties found in picking.

CONTOUR FOOLISH-LOOKING DATA?

In contouring, there is a question of what to do when the data looks foolish. You can use your judgment at the time and throw out, or bypass, the numbers that look like they don't really belong with the rest of the data. That will give you a smooth, geologically reasonable-looking map.

That is a sensible thing to do—unless the numbers are trying to tell you something. As in the exercises, a group of wrong-looking numbers, if all along one seismic line, may be saying that the line was mislocated on the map, or the section mis-labeled, or a different correction applied to that line, or something. If you had smoothed the contours, you would have made the map look better but might not have noticed the coincidence of all the bad points being on one line.

On the other hand, if you had gone ahead and contoured the data as if you believed it, funny kinks in the contours would have lined up along that line. Then you could have made the proper correction for the problem. The resulting map, too, would have looked smooth and geologically reasonable but would also be more correct, that is, a better representation of the subsurface, than the deliberately smoothed map.

Or the kinks might be, not on one line, but lined up fairly straight across several lines. Tight contours might indicate a fault, which would have been missed on the smoothed map.

There are other ways of handling this. Some interpreters may be able to see the lineups of kinks in the data before contouring. This is a skill that is difficult to acquire, though, and at its best, may not be as good as seeing the lineups in their effects on contours.

My recommendation to you, then, especially when you are a beginning interpreter, is to contour the data on the map. If the data is foolish, then contour it to look foolish. That makes the particular type of foolishness easy to see. Then look at the contoured map and take whatever steps are necessary to make things work out better. You may then decide to throw out some data, but at least you will have a better idea of why that was necessary. Then you can fix up your contouring.

Contour the data as it is to point out problems; then try to solve the problems; then re-contour.

CHECK YOUR CONTOURS

You should be optimistic in exploration, but you must also make strong efforts to be correct. Finding a "prospect" that was created by sloppiness is not so good. At all stages of your work, you need to check yourself. A point may appear high because a reflection time was misread or misplotted. To check this, look at the point on the section to see if the reflection bends up at that point.

When you contour, you should form a habit of always checking the contouring. Even better may be to get someone else to check your maps and you check that person's maps. This will prevent your assuming something is right because the decisions you made in contouring are still firmly in your mind.

Contour checking is particularly easy with the contouring method that doesn't run contours through shot points. As I explained, you just treat the space between two contours as a path and go down that path making sure that all the times found in the path are in, for instance, the 2540s, that is, 2540 through 2549. It is the shot point circles, of course, that must be in the path, not necessarily the numerals themselves. That is easy enough to keep track of, though.

It is convenient to do the checking on a map that you can mark on, a work map or an extra print of a drafted map. On it, you can keep track of which paths you have checked, by making pencil marks at the ends of the paths or lines down them. When all the paths are marked, the map is thoroughly checked.

When you find an error, mark it, preferably in a bright color, and go on checking. After the checking is finished, you can correct the errors. On a work map, you can correct them by just erasing and putting the new contours in their places. On a print of a drafted map, you can change the contours, preferably in color, so the draftsman can easily find the new contours to make the changes. In either case, it makes sense to check the map again after the changes are made.

This is a lot of checking, but it is worth the effort. Other people may later find things that are not correct in your work, but you should try to make all the corrections you can yourself, before they see it. So—

☐   Check the contouring on your work map.

☐   Check the contouring again after the map is drafted, to check the drafting of both contours and data.

That is an important point about the data. A map is drafted and needs to be checked before being put in a report or sent out to other offices. Someone could check every number individually. But in that tedious, repetitive kind of checking it's likely that a person will fail to find errors, just because the mind is lulled into not being very aware. If you check the contours, any number that is badly wrong will stand out. The badly wrong numbers are a lot more important to catch than the ones that are a little wrong. If someone checks all the numbers, fine, but after that the contours should be checked, too.

Notice that in checking final contours, you are checking a number of things. You can find errors that were made in:

☐   Your picking of the sections,

☐   Your timing of the picks,

☐   Your plotting of data on the work map,

☐   Your contouring of the work map,

☐   The draftsman's copying of the data,

☐   The draftsman's tracing of the contours.

Whatever may have slipped by any earlier checking is given one final check for reasonability.

Now, in Exercise 6–18, check the contours of an already contoured map.

There were errors on that map; I hope you found them. They are the kind that people often make in contouring.

Exercise 6–19 has another set of contours to check. This is an area with less relief, so the contours wander about more.

Kind of fun finding where somebody was wrong, isn't it? Keep enjoying it, especially when you're the one who was wrong.

Now for a contouring error that is more difficult to correct, do Exercise 6–20.

CONTOUR IT AGAIN

Contouring is rarely a unique, single-solution way to represent data. If the dip is steep and there is close seismic control, then one contouring of the data is about like another. But the economics of exploration does not usually permit the close control that is necessary to obtain this unambiguous contouring. Certainly in the reconnaissance stage, the close control would not be worth the cost. When a prospect is ready to be drilled, there is good reason to have sufficient control to make sure the prospect is worth drilling; but even when there is enough control to be sure that the prospect is a good one, there may be choices in just how it is to be contoured. There may, for instance, be two valid ways to contour, both of which make it a good prospect and point to the same spot to be drilled. In this case, it is not worth extra delay and expense just to answer an academic question about the contouring. If oil is discovered, the question will probably no longer be academic.

Anyway, you will not very often have enough control to assure you that there is only one way that the data can be contoured. But after working on the data and then on the contouring, it is easy to convince yourself that you have come up with the one right interpretation.

When you think that, your skepticism is at a dangerously low level. Then is a good time to get another copy of the map with the data on it and start contouring again. Do something different this time. Start from another part of the map or turn the map around, or make some assumption about the trends in the area, or connect the faults in a different way. Even better, get someone else to contour the data, preferably someone who hasn't seen your contoured map. Any of these different approaches to the contouring will probably bring out some things about the area that will surprise you. Surprises may point out flaws in what you thought were prospects or may bring out possible prospects you hadn't noticed.

One of those approaches, starting with an assumption about the trends, is opposed to part of the philosophy of exploration I recommend, that of letting the data determine the trends, of not having preconceived notions. However, now we are talking about a second interpretation, a search for alternatives. That is different. You can do all sorts of things to see what might be possible. And if some assumption makes things fit together much better, then that assumption has been to some extent verified by the data.

COMPUTER CONTOURING

Computer programs have been developed to do approximately the things people do when they contour. Why bother to learn to contour by hand, other than to understand what the computer is doing?

Using the programs calls for considerable effort to get the data ready. First, all the shot point locations must be put in a form that is meaningful to a computer, as X and Y coordinates, that is, horizontal and vertical distances from some point. Then those coordinates and the data to be contoured must be put in computer memory. Only after that is done can the program turn out a map.

All that effort isn't worthwhile for a map that doesn't take long to contour. Its main use is for great quantities of data. The largest quantities of data to be contoured are from airborne magnetometer and gravity surveys. The contouring of such voluminous data is routinely done by computer.

For seismic data, machine contouring is most likely to be worthwhile if the locations are already on computer magnetic tape and if the area is large or there are a number of horizons, intervals, etc., to be contoured. In offshore work, the locations are determined by computer and so are already on tape.

Machine contouring isn't yet in general use for seismic data. One reason for this is probably because of the delay inherent in having things done by a large computer. The computer itself is fast, but the large computers are too expensive to be allowed to be idle. There must be a routine of keeping different jobs prepared for the computer so it can run continuously. The contouring of a map might take some weeks from the time it is sent to a data processor until the time it is returned, complete. The delay is being reduced as computers become smaller and cheaper, so they don't have to be kept constantly busy.

Downtime of a data-processing minicomputer isn't a problem for an oil company that has one in its office. And personal computers are beginning to take over such tasks, with no more concern about the time they happen to not be in use than there now is about a typewriter. Some companies are now using personal computers to contour seismic data.

A computer contours a map in one way, a way that is determined by the program put into the computer. If it contours the same data again, with all the parameters in its program the same, it will produce the same contours again. A person contouring by hand can contour one way and then another, weighing the two against each other to see which makes better sense in the light of other factors, regional geology, the way a nearby area can be contoured, optimistic and pessimistic views of the area. Computer contouring also does not handle faulting as well as people do.

For now, most seismic data is contoured by hand. Your ability to contour by hand will be useful to you now for contouring, and when you do use computer contouring, it will enable you to understand the computer better and to get better use out of the computer than you would if you just accepted its contours without understanding.

Make Some Maps

We've gone over the ways to put data on a map and contour it. There are more aspects of mapping to be considered now. Maps of seismic time to a horizon are not the only maps used in seismic exploration. Also, there are things to know about putting data on maps legibly for presentation to other people.

OTHER KINDS OF MAPS

Seismic interpreters also make depth maps, time interval maps, maps of regional features, maps of seismic attributes, topographic maps. In some areas you may need to make several of these kinds of maps to properly evaluate a prospect.

DEPTH MAP

When we make a map of seismic time, it is intended to show the structure of a horizon in the subsurface. Obviously it doesn't show structure directly. Structure is a matter of depth, and the map is in travel time of sound waves. To make a map that is more truly related to the subsurface shapes, depths must be calculated from the times. The idea of converting the times to depths is so reasonable that it makes a person wonder why seismic maps aren't always produced in depth.

There are reasons for making the time maps. The times are read directly from the sections and so are immediately available for mapping. Because they are read from the seismic sections, they are the actual data obtained. Anything else derived from them must go through some conversion process, which may introduce errors to the data. The conversion of seismic data from time to depth is very subject to error.

To get depth from time, you need to use a velocity. If you are driving a car and wonder how far you have gone, not having noticed the distance reading when you started, what can you do? Look at your watch to see how long you've been driving; then estimate your average speed. Say you've traveled for one and a half hours at about 60 kilometers an hour. You've gone 90 kilometers.

This is very like the seismic situation. We can measure times very precisely but have only an estimate of velocity. The correctness of the fine measurements is partly thrown away by applying the estimate to it.

Velocity measurements made with check shots at wells are very accurate, but the wells are usually far apart and the velocities can change in a short distance. I mentioned only the check shots because a sonic log, although it gives fine detail, is not as correct overall as the check shots.

Velocity determined from seismic data, that is, from normal moveout, is not very accurate, and the deeper it is, the poorer it is. Even the reflectivity measurements used to make synthetic sonic logs add detail, but not overall correctness. If you convert your maps to depth, you are likely to have problems when you tie wells.

For these reasons, geophysicists generally make their maps in seismic reflection time. But then there can be a problem communicating with managers, who need to know about depth. A good practical solution to this problem is to leave the data in time but to label the contours with both times and estimated depths. Then people can look at a map and see about how deep the horizon is at any point on the map.

It is usually better to map in time rather than depth, but there will be occasions when, in spite of all the problems, you do need to map in depth. Velocity pullup or pulldown makes the time structure incorrect below the feature that causes the velocity effect. Some areas have so much change in velocity from place to place that the structure apparent on a time map is quite different from the structure measured in depth. In developing a field, there is more need for depth information than in exploration. Fortunately, as wells are drilled, there are more points to tie to and more velocity information.

Now that I've pointed out the problems and the need in some cases for depth maps, how can you best make them? Every area you will work is different, so I can't really tell you what will work best in a particular situation. You may have to try several approaches before finding one that seems to work well in the area.

The simplest and easiest way of converting to depth is by using one average velocity over a whole area. Multiplying time by velocity gives distance. But the sound went down and up, and you just want the distance down. You should use half the time or, to do the same thing, divide the whole thing by two. Or to make the calculation easier, use half the velocity. That is easiest of all, when you use the same velocity over an area. You have to divide by two just once, not once for every shot point.

The trouble with a uniform velocity, of course, is that you won't often encounter areas with velocities so nearly the same from place to place that one will be good enough for the whole area. And if you do have such an area, a time map is about as good as a depth map.

For a velocity that varies over an area, you can make a velocity map. Put the average velocity to the formation on the map at each well or at some points where you can get fairly good velocity information from normal moveout or at both, if you can make them fit together consistently. Contour the map. Because of the way you made it, it will fit all control points. In particular, it will fit the wells. Read a velocity from the map at each shot point at which you have read a time and use that velocity to convert that time to depth.

This depth map will fit all the existing wells exactly, so you won't have any mis-ties. But by making it fit each well you have been forcing the data. If another well is drilled, the data may not tie it. A fairly good test of how reliable the velocity map is for new points is in its complexity. If it is highly contorted, going up for one well and down for the next, it probably won't predict the velocity at other points very well. If it is smooth, generally increasing in one direction, then it is likely that a new well would fit it closely.

For a little practice in making this sort of map, work Exercise 7–1. Don't forget to use half the time or velocity. Forgetting to divide by two is a very common mistake in geophysics.

Notice how different your contoured time and depth maps are. Notice especially whether you might have recommended drilling a well on the time map and compare it with the recommendation you might make based on the depth map. This isn't a common velocity situation; but where it does exist, you can see that correcting for the velocity variation is vital to finding the high points in the area.

You should also check the velocity map for similarity to a reflection time map of a horizon. If the velocity map looks a lot like the structure map, with its highs and lows in the same places, then the velocities are probably changing with depth of the horizon. In this case, you will probably be better off not using the velocity map to pick velocities from, but using it only to tell you that there is a smooth relationship between depth and velocity. On grid paper, you can then plot a number of the reflection times to one horizon against velocities to that horizon at those shot points. If the scattering of points is such that you can draw a smooth curve through them, with the points clustered fairly closely around it, then the curve may be your best velocity function for the horizon. For any time on the graph, you can read a velocity that could be used to convert that time to depth. But to make it easier, do the converting for the times and draw a new curve, this one of time against depth. Then to convert a map, just look up the time on the curve and read depth. If the map has a lot of points to be converted, you may find it even easier to put the time-depth relationship in the form of a table of numbers rather than a graph. It will take time to make the tabulation, but after it is made, looking up depths will be quicker and more consistent. To make the tabulation, you wouldn't read each time from the curve, as you couldn't very well do that and keep the depths increasing smoothly. Instead, you would read the depths for a few times and interpolate arithmetically between them.

You may notice that I was describing what to do for only one horizon, although several horizons may be picked on a section. For more than one horizon, you can do as you would with only one, but with one difference. In doing one horizon, we used average velocities from the datum plane down to it. If we also did that for a second, deeper, horizon, there could be a conflict. If in some part of the area the two horizons come close together, the two independent depth conversions might happen to give the deeper one a shallower depth than the upper one. The horizons wouldn't be far out of position, but the effect is worrisome and makes it easy for people who don't understand exactly what you have done to lose faith in the interpretation.

You can avoid this problem and also have slightly more correct measurements. Measure the times and velocities from the surface to your uppermost horizon. Then for the next horizon, determine the times and velocities from the upper horizon down to the next one. These are interval times and interval velocities. Make a curve or a table for this interval below the top horizon. Then, to convert the second horizon to depth, do the conversion in two steps. Look up the time to the first horizon and get the depth to it. Then look up the time between the two horizons in the second plot or table to get that thickness. Add it to the first depth to get the depth from the surface to the second horizon. For a third, use the interval from the second down to the third, and so on.

This "peel-off" method is quite useful. A version of it could also be used if the velocities did not seem to vary with thicknesses of the zones between reflections. Large lithologic changes might make velocities vary with the part of the area more than with the thickness. Then you would use whatever relationship did seem to apply, perhaps making velocity maps of the various zones and plotting interval thicknesses to add together at the shot points.

If you use a method that does not force the reflections to tie the wells, then after determining depths, you may need to concern yourself with the problem of why some of your ties are not good. This, of course, is the same thing you must do in making time maps and in the picking itself. The mis-ties on a depth map just make the problem more obvious.

TIME INTERVAL MAP

An isopach map is a map of thickness of some zone. A geologist may make such a map by picking two horizons on well logs, subtracting the depth of one from that of the other, putting the differences on a map, and contouring. The contoured map then shows how the thickness of the zone changes from place to place. There are refinements, like calculating true thickness where there is dip, rather than just using vertical distance.

An isopach shows differences in amount of material deposited. For this purpose, the isopach is usually made over a rather thin zone, maybe a single formation.

A similar map can be made from seismic data. The easiest and most direct way to do it is to subtract the reflection times of one horizon from those of another and map the difference in time. This is not an isopach, as it is in time intervals rather than thicknesses. But to make an isopach from seismic data requires converting two horizons from time to depth, with the errors of both included in the resulting map. Like depth maps, these isopachs may sometimes be necessary but are not advisable unless they are truly needed.

There is a bit of confusion between two parts of the world as to what to call a map of time differences like this. In the U.S., an ordinary map of reflection times from the datum plane is called a time map, or time structure map. In Europe, such a map is often called an isochron map, from Greek *iso*, "same," and *chron*, "time." A contour on it is a line connecting points having the same time. Other similar words in the sciences are isobar, for contours of the same pressure, and isotherm, same temperature. Americans, less familiar with Greek roots, don't call a normal time map an isochron. Instead they thought up isochron by supposed analogy with isopach and now use it to mean a map with contours representing equal time intervals. I was one of them but have reformed. Other Americans, still missing the same point, use a two-language word "isotime." The confusion from either of these terms can lead to communication problems that are potentially expensive. To avoid the problem, I call these maps time interval maps. It's wordy, but at least it won't cause a vast misunderstanding in some transatlantic telephone call.

Enough of linguistics. A time interval map shows the thickness of the zone between two horizons. Or, like a section made by flattening a horizon, you can think of it as showing paleostructure, the structure of the lower horizon at the time the upper horizon was deposited. That is, the time interval map is the map you would get if you flattened a horizon on a number of sections to show paleostructure and then picked some deeper horizon on those sections and mapped it. Like the flattening to remove glacial drift, the upper horizon can be very shallow, and the interval map can be thought of as representing the structure of the deeper horizon with the near-surface problems corrected.

In Exercise 7–2, make a time interval map from two existing seismic maps.

Compare the map you have made with the horizon maps, particularly the map of the deeper horizon, to see the differences between present structure and paleostructure.

In flattening to show paleostructure, faulting creates a problem, the same problem that faulting caused in flattening a section. Within the horizontal distance occupied by the fault, the thickness doesn't show the paleostructure, but just shows that there is a fault at that place. About all you can do is to mark the boundaries of the fault on the interval map. If the horizons are far apart, the horizontal distance between the point at which the fault cuts one horizon and the place where it cuts the other can be pretty large. You can lose a considerable amount of the area of your map to faulting.

There are two ways to make a time interval map. One is as described, by subtracting the times at the shot points. The other is quicker but is dependent on the choices that have already been made in contouring. To do it, contour both horizon maps. Overlay one on the other, matching locations carefully. Then subtract, not at shot points, but wherever a contour on one horizon crosses a contour on the other. This subtracting is easier, as the times of the contours are round numbers. But it gives a new scattering of points, not at the points where measurements were made. And it includes whatever errors of judgment were made in contouring, adding together the errors from both maps.

Do Exercise 7–3, using the same maps you used in the previous exercise to make an interval map by subtracting at contour intersections.

Compare the map you have just made with the one you made on the exercise before this, to see what differences the two methods make.

SEISMIC ATTRIBUTE MAP

The various characteristics of seismic data that can be displayed on sections can also be put on maps. Frequency, polarity, amplitude, whatever qualities may seem particularly useful for an area or prospect, can be obtained in numerical form, plotted at shot points, and contoured. It may then be useful to overlay a transparent copy of that map on a structure map to see where both structure and stratigraphy are favorable for the accumulation of oil.

FEATURES MAP

Some geophysicists like to look over the sections before doing the detailed working of loops and just pick out the major features—the regional faults, major structural trends, subcrop lines, etc.—and put them on a map. This is a features map, which can be used as a guide and a way to keep from getting lost in detail when picking reflections around loops.

TOPOGRAPHIC MAP

There is often a need for a map of the topography in a seismic area. On land, the surveying has determined elevations at the shot points, primarily for use by the data processors in making datum plane corrections. Offshore, a depth recorder has supplied water depths, which are elevations of the sea bed.

On land especially there is reason to map and contour the elevations. If there are not good published maps of topography, or air or space photos, then a topographic (or topo) map made from the survey data may be helpful to a later seismic operation in the area or to a rig move for drilling a prospect. Even if there are good maps available from other sources, the fact that the shot points were specifically surveyed makes the seismic surveyor's data worth mapping. Another reason for mapping elevations is to check interpreted maps. If near-surface corrections were incorrect—too large or too small—then a seismic map of a flat layer of rock may look like the topography or like a mirror image of the topography. If the layer isn't flat, and it usually isn't, the situation is more complicated, but the map can still be used to judge the near-surface corrections.

For these two purposes, there are separate philosophies of contouring. For a map to use in getting around, all the features should be included. The surveyor, if available, can be asked about streams and hills that are not on the seismic lines, or asked to contour the map, in part from memory of these features. Or published maps can be used to add information. This will produce a map that is a good representation of the surface of the ground.

But for a topo map to be of most use in deciding whether the shape of the ground is influencing the subsurface map, the extra features are a detriment. This kind of topographic map should be contoured from data only at shot points and contoured the same way a map of seismic data would be contoured. It and the seismic map will be more similar and will better show the similarities that indicate a near-surface or velocity problem.

SITUATION MAP

There are some non-seismic types of information that are often useful on a map:

☐ Topographic features—coasts, rivers, hills, etc.;

☐ Political boundaries;

☐ Land ownership, lease, concession outlines;

    ☐   Artificial physical features—power lines, towns, cultivation, etc.;

    ☐   Survey markers—bench marks, wind socks, etc.

There may be so much of this extraneous information that it would obscure the seismic information. Some companies handle that problem by putting all the extra information on a separate map, a situation map. Then the seismic data can be on one map with only a minimum of other information, and all the rest can be handy on another map.

COMBINING DATA

There are a number of different parameters that can be shown on a contoured map, and it may be useful to compare two or more of them. You can map:

    ☐   The seismic time to a reflection;

    ☐   Depth obtained from the seismic time;

    ☐   Interval time between two horizons;

    ☐   A formation's zero edge;

    ☐   Thickness of a reef in seismic time;

    ☐   Reflection quality or other character variation;

    ☐   Reflection amplitude;

    ☐   Polarity of the reflection;

    ☐   Frequency of the reflection;

    ☐   Bright, dim, flat spots;

    ☐   Situation map information.

There are several ways of displaying two or more kinds of data so they can be compared readily.

    ☐   Of course, it is always possible to have the different sets of data on separate maps and look at them side by side.

    ☐   A variant of this is to lay one over the other, carefully placed in register. Then, holding the part farther from you firmly down, repeatedly raise and lower the part near you. The eye, or the mind, retains one image a bit while seeing the other. Some people have developed this technique to a high degree of skill.

☐ You can fold one map and lay the folded edge at the same position on the other map. You are then seeing a map made up of parts of the two maps. You can re-fold to match the maps at other places.

☐ If a light table is available, or even a window, one map can be laid over the other and both seen together by transmitted light.

☐ One map can be made on transparent material and laid over the other. This is often used in reports, with the two bound together.

All those methods give a choice of seeing both or only one. But the different kinds of information can also be combined on the same map in different ways.

☐ They can be drafted on the same transparency, so they appear on all prints. If similar things, like two sets of contours, are on the same map, then it helps for them to be visually different, with one heavier, or dashed, or with one in ink and the other in pencil. Different-looking data, like straight-line lease boundaries and curved contours, do not interfere much with each other.

☐ One can be omitted from the transparency and added later to each print in color. However, it may be better to have it drafted, maybe lightly, on the original and then go over it in color on prints if desired. Then it isn't necessary to hunt up an already colored copy to trace the colored lines from.

We have discussed various forms of maps used in seismic interpretation. Let's go further into the subject of mapping, taking up the things an interpreter needs to know about making the maps legible and presentable.

DRESS UP THE MAPS

Drafting is the part of an office operation that makes your interpretation look professional. An area can be thoroughly interpreted and oil discovered—all from undrafted work maps. So why do we bother with drafting? Why not just interpret and find the oil?

The very first point is that most of the "undrafted" work maps were drafted before they became work maps. The shot point locations had to be put on a map with legal subdivisions, some coordinate system, topographic features, information enough to make clear where the shot points are on the earth.

When oil is found directly from work maps, the work is usually done by a small company. Only a few people are involved, and the interpreter is in close contact with the person who makes drilling decisions. The work map is not only shown to that person after it is complete, but may frequently be discussed with the person while it is being made. Sometimes the interpreter and the decision maker may even be the same person.

In a large organization, or when the interpreter is stationed in a location remote from the head office, there is not the day-to-day communication of the small one-location office. When maps are shown to management, they may be sent by mail or presented in an out-of-town meeting. There may not be a chance the next day for them to ask what something means. The way to handle this is to make things as clear and unambiguous as possible. Lettering needs to be very legible, contours clear. The map needs explanatory notations about prosaic things like type of map projection, scale, and date, which aren't necessary on the work map. The map also needs to be reproducible, so copies can be distributed to different people who may be involved in the prospect in one way or another.

These are reasons more vital to the use of the map than just prettying up. But making it look nice is of value, too. When an interpretation is shown to someone not immediately involved in it, a neatly drafted map makes a pleasing impression just because it does look good. It also indicates that the interpretation is one that was thought worth drafting. The finished product obviously wasn't just thrown together the night before, to have some kind of map to show. Drafting is an important part of presenting your interpretation to people who are able to do something about it, of the salesmanship of interpreting. And that is essential. A superb interpretation that never goes beyond the interpreter's mind and desk drawer is of value only as a practice exercise in interpretation.

After doing the work of making an interpretation, get it drafted.

DEALING WITH DRAFTSMEN

The way to have the drafting done well is to take the work to the draftsman in plenty of time; explain carefully what you need; consider the draftsman's ideas about what may be better, quicker, or clearer; look in on the work occasionally while it is being done, without interfering too much; and don't change your mind after you've turned the work over for drafting. That's how to get it done well. A number of draftsmen have asked me why we can't do it like that, but in different words: "Why didn't you bring it to me sooner?," "Do you always have to be in a rush?," "Can't you people make up your minds?"

It might be pleasant to live in such a peaceful, unhurried world. But in a competitive business, one with very expensive operations waiting on interpretations, there isn't much way to avoid being in a perennial rush, having last-minute changes, and all that. Perhaps a rig finishes a well sooner than expected. Standby time costs tens of thousands of dollars a day. Nobody wants to pay at that rate for a geophysicist to be leisurely and orderly in getting things ready for drafting.

On the other hand, you need to do your best to work well with the drafting department. They catch the very end of the rush jobs. They can't even start work till the geophysicists and other technical people are ready. So management frequently hears "Waiting on drafting." Of course. The geophysicist aimed at the deadline and got through just in time, if you assume the drafting could be done in no time at all. Maybe the geophysicists and the geologists and the landmen got through just in time, and all handed their work to the draftsmen at once, to pretty up for the meeting or other deadline. It can't be done that way, so it won't all be done. The people with the most clout, or who scream the most, will get their maps rushed out, and the other people's work will not be done. And everybody's mad at the draftsmen!

To minimize this problem, there are several things you need to work on. I do mean work on, not just be aware of. This is a part of the work of geophysics and requires your time and effort.

Give the work to the draftsmen in time, if possible. Sometimes you can. In any case, don't wait till the last minute just because there doesn't seem to be any hurry. Or, when there is an unavoidable rush job coming up, give the draftsmen warning. Maybe they can work on some part of your mapping before you have the rest finished. Maybe they can be putting the numbers on a transparency while you are contouring a work print. Then, at the last minute, all they have to do is add the contours.

Explain to the draftsman what you need. Many people tend to assume that draftsmen know how to make maps. Well, they do; but if you just say, "Put this on a map for me," you may find out, after some work has been done on it, that you had more special, unusual, requirements than you thought. It's a matter of intellectual laziness. It takes a lot of thinking to work out, and then explain, details like how the data should be displayed, which contours to make heavier, etc.

Consider their suggestions. They have been preparing data for different people and different requirements. They probably have some ideas that are better than yours. A good discussion of what you have, and what needs to be shown on a map, will bring about a joint judgment.

Look in on the work often enough to prevent some misunderstanding from being the cause of a lot of wasted effort and maybe missing a deadline. But don't get in the way of the work and don't interfere in the decisions that the draftsman can make as well as you.

Try not to change your mind after the drafting is started. When you can, have everything settled before turning the mapping over for drafting. But in a rush, you both have to make the best of it. For instance, new data may come in, and things may then just have to be changed. Or, to give them a head start, you may have got them to put the data on while you contoured, and the contouring may have pointed out discrepancies that made you change some of the data.

Another way of cutting down the rush, and the expense of drafting, is to use some tricks to reduce the amount of drafting that needs to be done.

DRAFTING—HOW TO AVOID IT

Drafting can sometimes be avoided by the use of photographic techniques. Drafting, as a separate step, can be reduced by being a moderately good draftsman yourself and combining this with photography. There are some tricks for getting lettering on maps quickly but neatly.

We'll start with the last item. If considerable text, explanation, etc., is to be put on a map, lettering can be added much more rapidly by typing than with a guide and technical pen. Cut a piece of whatever the map is on, film, tracing paper, sepia paper, etc., to fit in a typewriter. Put it face up, backed with a sheet of reversed carbon paper, that is, with the carbon side up, so the carbon image will be on the back of the film or whatever. Special yellow, orange, or red carbon paper is made for this purpose, but ordinary typewriter carbon paper is just as effective. Then type on the transparency. There will be type on the front of it from the typewriter ribbon and on the back from the carbon paper. This double layer of type is more opaque than just one layer and so can be reproduced better by the ammonia-developed diazo process. Then the film can be inserted in the map, by cutting out a piece of the map and taping the typed piece in its place.

There is a special trick to taping a piece into a map. The two parts must lie very flat before taping begins, to avoid buckle when the spliced map is run through the rollers of a printing machine. If the map does buckle in printing, some of the map may be buckled up out of close contact with the sensitive paper, so the image of that part is out of focus and therefore blurred on the copy. Film is the best substance for being free of stretch and therefore free of buckle. A paper map may buckle some no matter how careful you are with it. Although a piece can be successfully spliced into the interior of a map, the four sides offer opportunities for buckling. Reducing the number of splices is the best way to reduce those chances. If the added piece can be in a corner, so only two splices are made, the map will tend to be smoother. This will probably call for re-drafting part of the border of the map, but that is not much work. Still better is adding the piece all along one edge of the map, so only one long straight splice is necessary.

Incidentally, an image of the tape used to make the splice may be visible on a final print, but if the tape was applied neatly it will not look bad. The tape used is usually intended to be as transparent as possible, to reduce its visibility on maps. Alternatively, though, an opaque tape can be used. It shows as a wide dark border setting off the typed notation and serves to call attention to it. Like contrasting patches on blue jeans, it makes the necessary appear to be intentional.

A trick for quickly adding a small notation involves applying the type to a strip of tape. Type the note on a piece of film with reversed carbon paper under the film in the typewriter. Then remove the film from the typewriter, lay it face down on the table, and stick a strip of transparent tape on the back, on the carbon image of the words. Pull the tape off the film and put it on the back of the map. Most of the carbon will have come with it, so there will be lettering on the map. With it on the back, it will be close to the printing paper when the map is printed, which will help make the image sharp. Also, the type is on the sticky side of the tape, between tape and map, so it is protected from abrasion in handling and printing.

Neither of the above two methods makes an image as opaque as India ink, so the type looks relatively weak. But it is quick, it doesn't take much of the draftsman's time when the rush is on, and it does look surprisingly professional.

A similar technique makes use of an office copier. Type the note on ordinary typing paper. Copy it on the regular copier, but with a sheet of film, sepia paper, or tracing paper in the copier. The transparent sheet can then be spliced into the map. If the copier is working well that day, the image can be quite opaque.

A possible problem with adding any tape to a map is that, in the heat of the printing machine, the glue on the tape may soften and allow the tape to slide off. This isn't usually a problem but may become one when many prints are being made. Then the steady use of the machine gets it unusually hot.

IF YOU NEED TO DO IT YOURSELF

There may be times when there is no way to get your map in shape but to do it yourself. Be prepared to, before the time comes. It may make a big difference in your presentations and therefore in your career.

USE YOUR WORK MAP

You want your maps ready when you need them, not some other time, after the opportunity to present them is lost. One way to be sure they are ready is to do the drafting yourself, on the work map. You can make your own life easier by being a sort of pinch-hit draftsman, mostly by having a little skill at hand lettering. Even if you are only fair, there are situations in which your results can come out looking pretty good. Mostly these involve photographic reduction, the greatest drafting improver ever. Arrange for your work map to be at a larger scale than you want your final map to be, maybe twice the scale. In interpreting, put data and contours on the map in ink, or ball point, or even nice erasable pencil. Then you can have it shot down to the final scale, and not only will the drafting look better because it is smaller, but the marks will all be in the same photographic medium.

This won't work if the double-scale map is too big to handle conveniently, or if the drafting isn't good enough even after reduction, or if your company insists on higher standards than hand lettering. But it is a method that can sometimes save the day.

One time a friend of mine had reworked a large area, incorporating the existing interpretations of a number of smaller areas into one map. Putting it all together produced a rather huge map. He had worked on a paper print and had put data on with a ball-point pen, so it would stay while he did the erase and draw, erase and draw that is often part of hand contouring. The contours were in pencil. The map was very useful. We got a lot of exploration value out of looking at his paper work map with the ball point and pencil on it. We decided we needed to get it drafted, make prints, and incorporate it in a report. But there was so much data on it that it would take a lot of the draftsmen's time to put the information on a new map, they were already very busy, and there just wasn't a good way to get it drafted.

Then I realized that his ball-point lettering was neat and uniform and large enough to still be legible after some reduction. He went over the contours, again with pencil, but this time with a softer one, made them thicker and heavier. We sent it to a reproduction shop and had it reduced to half scale on positive film. Prints from that film looked very good, very final and professional. The report was ready in time for a meeting; no panic had been caused; no errors had been made in copying his data onto another map. Huge success.

If you think this situation might arise, there are some precautions you can take. You shouldn't slow up you work much, just on the off chance that the work map itself might be preserved for posterity, but you can use some simple precautions that won't take much extra effort. It you use a ball-point pen for data, use one that writes black. It prints or photographs a lot better than blue. If you use pencil, use a fairly hard lead so it won't smudge. Then if you later need to copy the map, you can go over the pencil lines with softer pencil or with ink.

The main precaution is this business of hand lettering. No matter what your work map is used for, it will help if the lettering is fairly legible. It will even help you read your own numbers. Some practice will make your lettering more legible. Using guidelines at the top and bottom will get your hand in the habit of making the characters uniform in height. Uniformity is a big step. It alone makes even poor lettering look fairly good. Some draftsmen draw guidelines lightly, with hard pencil, and leave them on the map. If the lines are light enough, they don't appear on prints.

For style, you can imitate the form of the mechanical lettering on maps, which is designed especially for legibility. For instance, the number eight is made of two ovals. The longhand or ice-skating figure eight is not as legible from a distance. The two ovals make for better distinction from other figures.

A company I worked for one time used a printed form for lettering practice. The form had model lettering examples to copy and guidelines already drawn. The company even for a time required interpreters to fill the forms in regularly. It was time-consuming, and we grumbled a lot, but it did improve our lettering. I don't see much point in becoming highly skilled in lettering, at great effort, but a moderate amount of effort—say as doodling, when you want to rest the old brain—can pay off quite well.

PRINTED GRIDS

Another simplifying device for drafting is the generous use of printed grids. Printed grids can be bought on several types of surface. Opaque paper, transparent paper, transparent film are available with grids of various line spacings for plotting data. The grids are also in several colors: red, green, black, blue. Some, like the red and black, are made to reproduce well by diazo. One, a special light blue, is made to not reproduce by the diazo process.

There are two ways to make your drafting easier and neater with these materials. One is to draw your map or whatever on the transparent material that has the light blue rulings. It is referred to as fade-out or no-print paper or film. The other way is to tape a grid of any kind, on any drafting material, in any color, under a transparent sheet and then do the work on the transparent sheet. Or tape the grid onto your desk top and lay or tape a transparent sheet over it to work on.

The advantage of the grid, used either way, is that no special draftsman's skills are needed to draw lines that are parallel or perpendicular to one another. If you have to somehow put your own map together, you can use the grid lines to position border, title block, latitude and longitude, other lines.

With the grid taped to your desk, you can draw the vertical and horizontal lines, and then turn the map to use the grid as guidelines for hand lettering that is to be at some angle other than horizontal and for drawing other lines.

PERMANENT PENCIL

During interpretation, you may need to make some quick work maps. Pencil is easier to work with than ink, as the pencil can be erased when you change your mind. But if you put data on with pencil and then contour, things don't work so well. Unless you are a lot more skilled than I am, contouring is a process of repeated drawing and erasing. It's a lot of trouble to avoid erasing data while erasing contours. It is better to have the data in ink, diazo, etc., something not so erasable.

There are two handy ways to have both the convenience of pencil and the permanence of ink. One is to put the data on in pencil, copy that map on the office copier, and then contour the copy. You may have to have your map in small sheets and tape them together, but the method is quick and quite satisfactory for work maps.

The second way is to put the data on one sheet in pencil, tape a transparent sheet over it, then contour on the second sheet. This can go through more steps, too. If you don't have a blank map to work on but do have something that shows shot point locations, like a map contoured on another horizon, you can trace the locations onto a transparent sheet with pencil. Then tape another transparent sheet over the first one; put the pencil data on the second sheet and the contours on a third. Then, if you have data on another horizon to be worked, the first shot point tracing can be used again.

When using this transparency overlay system, you need some way to avoid mismatching the sheets after they get separated. Make it a habit to always put at least two match points, preferably near corners, on each sheet. Then, if you take them apart for any reason, you know just how they go back together. The special points are better than trying to match some shot points. It's too easy to forget just which shot points you are matching.

An advantage of transparent overlays is that you don't need any reproduction equipment. If you are going out in the field and may need to do some mapping there, just take a supply of tracing paper or film, some tape, and some pencils. Cut sheets are better than a roll of paper or film. Rolls are clumsy to carry and easy to leave behind on an airplane or in a hotel room or office. Small cut sheets will go in a briefcase, traveling bag, duffel bag, or cardboard box. If you feel you must have a large map, then either tape small sheets together or take large sheets and carry them folded.

A disadvantage of the transparent overlay system is that you usually won't put extra information on it, like streams or property lines. But when you show your map to other people, they'll want to know where some feature is. Keep another map handy to lay yours over and, if your map is referred to much by others, get your information drafted onto another map soon.

SPLICE PAPER OR FILM

There are occasions when two pieces of film or paper, with overlapping images, need to be joined together. Parts of two base maps may be put together to form a single map of a prospect that overlaps the boundary between the maps. A xerographic copy of a seismic section, when made on the office copier, must be made in pieces, which are then assembled into a long section. A well log may be copied similarly. Anyway, there are many situations that require splicing, and many that will be made more convenient and quick if you have the skill to splice quickly and accurately.

To make a good splice, start with two overlapping pieces of a section or map. Lay them on a surface that will withstand a cutting edge. If your desk happens to have a glass cover, that's perfect. Sit there and be comfortable. Otherwise, go to a tracing table (for the glass, not the light) or have a handy piece of wood or glass to lay on your desk. Put one print down, face up. Tape it to the surface. Put the other down, also face up, overlapping so the edge of it exactly matches the same part of the image on the first print. Tape it down.

Lay a straightedge, preferably a metal one so you won't cut it, a little back from the matched edge, on the overlap. Hold it firmly with your fingers wide apart. Cut along it with a sharp knife. The type of knife with the disposable, snap-off blades is good. Go over the cut with the blade until you are sure both layers are cut all along.

Remove the scraps of paper. Tack the pieces to be spliced together with bits of tape. Undo the tape that holds them down. Carefully turn them over. Tape the splice on the back. That's it, except for trimming edges and maybe removing the bits of tape from the front.

In taping the splice, don't stretch the tape. It seems so neat to pull it tight as you place it in position. But tight tape will later draw up to its normal, shorter length, crinkling the paper.

There are reasons for putting the tape on the back rather than the front. Tape on the front of a section causes problems. Sure, some tape has a matte surface so it can be written on, but that surface doesn't have the same texture, absorbency, etc., as the surface of the section or map. Ink or pencil on it is darker or lighter and is more (or less) subject to rubbing off. The tape was designed to be written on primarily for books, sheet music, newspapers, and the like. They are printed on both sides. There isn't a blank back to be taped. The matte surface is far better for them than slick tape that won't take pencil or ink at all.

DRAFTSMAN'S TECHNIQUES

In addition to being able to turn out legible, sometimes even reproducible, work maps, it is worth your while to know something of what the draftsman does with your data, on the regular professional map and even, in an emergency, to be able to do a bit of that work yourself. You may sometime need some drafting done when the draftsmen are off duty, maybe in the evening or on the weekend or far away, as when you are presenting your work in another city. It may then make a large difference to you to be able to add one number or change one contour on your map, without ruining the appearance of the map.

PENS AND GUIDES

For data to be clear and reproducible, a draftsman usually puts it on a map in India ink, the blackest, most opaque kind of ink readily available. India ink has particles of carbon suspended in it. They make up the black part that stays on the paper or film after the liquid part of the ink has evaporated. The particles also clog pens, so a special kind of pen is necessary to handle India ink, a technical pen. It is sealed tightly to prevent drying when not in use. The ink feeds through a capillary tube. There are pens with different diameters of tubes for different line widths. To keep the tube from clogging, a needle in the tube, with a weight attached, can be moved back and forth by shaking the pen. A draftsman keeps pens in working order by shaking during use, capping when not in use, cleaning occasionally. This kind of pen, with a straight-edge, can be used to draw solid black lines for map borders and the like.

A technical pen can also be used for mechanical lettering, using a plastic guide with letters and numbers inscribed on it and a holder for the pen, to allow it to trace out similar lettering on paper or film. To do this mechanical lettering rapidly and well takes considerable practice, so it is best left to the professional draftsman. But an occasional few characters can be added or changed by an amateur.

RUB-ON LETTERING

A system that is versatile in lettering styles and is easy to use is rub-on lettering. Paper or plastic sheets are sold with letters, numbers, designs on the under sides. The image of the lettering is held to the sheet by wax or something like it. When the sheet is laid on a drafting surface and a character is rubbed, the character sticks to that surface, pulling off of the sheet. The characters can be more intricate than would be practical for mechanical lettering. They can have serifs, curlicues, etc.

This type of lettering is easy for an amateur to use, but you need some skill to get the characters transferred whole and all the characters of a word or number in a straight line. A disadvantage is that, when all the copies of a specific character on a sheet have been used, another sheet is needed. An office should keep enough stock ahead to avoid having to have someone rush out to buy one more sheet to finish a project, but not so much that the sheets get several years old and lose their transferability.

TRIANGLE SLIDING

One of the skills of a draftsman that is fairly easy to learn, and quite useful, is sliding triangles to make parallel lines. Lay a triangle (a set-square) with its farther edge along a line on the map. Then place another triangle against it, on the side near you. Hold the second triangle firmly down against the map so it won't slide and then slide the first along the second. Then hold the first triangle down and draw a line along its far edge. This line is parallel to the line you first laid it along.

This technique is useful in many ways—using guidelines for hand lettering, drawing special-purpose grid paper, aligning shot point numbers or data uniformly, etc. The main skill needed is holding the proper triangle down. The triangle to be held down is first one and then the other, so it is easy to get confused and let the wrong one slip.

SCALES AND STRAIGHTEDGES

For noncritical measurement, we may mark a distance with a ruler and then draw a line along its edge. But for drawing very precise lines, it is necessary to do the measuring and the drawing with two separate instruments, a scale and a straightedge.

A scale has the markings placed so they can be read **at** the drafting surface, not the thickness of a ruler above it. The scale may be very thin or may be transparent with the marks on its underside or have a triangular cross section or a bevel to bring the edge of the top surface down to the base. It is made of a material that does not stretch or shrink much with changes of temperature or humidity. For precision, the draftsman looks straight down from above the scale to eliminate parallax, the misreading caused by viewing from an angle.

A straightedge doesn't need markings, although there may be some on it for ruler-type convenience. It is made of a material that can have a very straight edge, usually metal or plastic. Stretch and shrink don't matter, as long as they don't warp the edge. The working edge may be beveled to a thickness that is convenient for running a pen or pencil along it. After measurements have been made with a scale, a line can be drawn between measured points with a straightedge. Again, the draftsman puts an eye directly above the edge and about midway between the ends of the line to be drawn. But the edge is not placed on the measured points. Rather, the pen or pencil is placed on a point and the straightedge pushed into contact with it; then the same thing is done at the other point and the first point checked to make sure that end of the straightedge hasn't moved. Then the line is drawn through the points.

A pencil line couldn't be drawn through the points with the straightedge closer than half the width of the pencil point. The same is true for a pen, but there is an added problem with it. If a line is drawn with a pen tightly against a flat straightedge, some ink is likely to run under the straightedge, drawn into the thin space between it and the drafting surface by capillarity. So the pen must be angled slightly away from the edge, or, better, the straightedge is made with the underside of the edge cut away, so the pen can be held upright without the capillary space touching it. If the edge is not beveled on the underside, the same effect can be achieved by putting some tape on that side, a little way back from the edge. There is a difference between straightedges made for different uses. A straightedge made for pencil is either not beveled or only the upper edge is beveled. One for ink is beveled a little on the under surface, and sometimes also on the upper surface.

Work Exercise 7–4, drawing lines exactly through some points.

The exercise was easy to do carelessly, but if you did it carefully enough to run a line through two points and not let it bend, you probably learned some technique that can be useful to you.

This has been a minimal exposure to drafting, an amount that you probably will need in seismic interpreting. When you are able to map the seismic data—time the picks, make the maps, and contour them— you have a start in interpreting the subsurface and finding some oil. We can now go on to the interpreting problems of geologically different areas.

Interpret Subsurface Features

We have discussed interpretation in general and are now ready to take up some specific types of features. Each geological situation has its own peculiarities that become familiar to an interpreter working with it. In working the area, the interpreter develops some specialized techniques to deal with these peculiarities.

OIL AND GAS TRAPS

There are two requirements for oil or gas to be trapped in the subsurface. First, there must be porous rock that has a local high point. The rock has to be porous, or there wouldn't be any holes in it for the hydrocarbons to occupy. Porous rock is mostly filled with salt water, and the oil and gas are lighter than the water. The place must be the locally highest, or the hydrocarbons would float on up to a higher place.

The second requirement is that the rock immediately above the porous rock must be impermeable. Otherwise, the higher rock would be the locally high porous rock, and the oil and gas would float up into it. This impermeable rock is the seal of the trap. For a rock to be permeable, it must be porous.

Let's think about the situation a bit. The fluids don't move very freely through any rock but make their way gradually from pore space to pore space. Solids can't make the journey, as they get stuck in the spaces, so people have a saying about water wells and springs to the effect that water purifies itself in some distance. The solids get left behind, so the water is cleaned of them. However, the liquid can still contain minerals that are actually dissolved. The material in solution can make the trip in the liquid but may be precipitated out where a different chemical or a different temperature is encountered.

For rock that is deep underground to be porous, it must be hard enough to withstand the weight of the overburden. Soft, mushy rocks are compressed until the fluids are squeezed out of them. The soft rocks can be pushed around somewhat like liquids, forming salt domes, shale diapirs, etc. After the pores in soft rocks have been mashed to nearly nothing, these rocks are good seals, with no room for fluids to move through them. The hard rocks resist the pressure, but when they do give, they are more likely to break.

ANTICLINES

The simplest type of feature to interpret seismically is an anticline, where rock layers are bent upward and, further on, back downward. Then, if one of the layers is porous, the anticline is a high place where oil and gas, floating on top of the water in the rock, can accumulate.

Anticlines are the type of oil trap most sought through the years of development of seismic exploration and the type most often found. Most anticlines are not very subtle. Even before there were seismic sections, reflections on unprocessed records of single shots could be picked as they went higher and then lower in crossing an anticline. The high points could be shown in three dimensions on maps. An anticline is the type of potentially productive feature that seismic exploration is best fitted to find.

With a seismic section, it is often easy to recognize an anticline. Just look for a place where the reflections dip down both ways from a high point. That is an anticline, but not necessarily an oil trap. To form a trap, the anticline must be closed, with dip downward in all directions, not just the two directions on the line, to prevent the escape of the oil.

If you happen to have another line that crosses the first one at a right angle, crossing somewhere high on the structure, you can gain more information about the anticline. If the reflections on the cross line just dip in one direction, then the closed high point, if there is one, is somewhere beyond the high end of the cross line. Or there may be evidence of an anticline on the cross line but with the high point some distance from the high on the first line. Fine, the structure is likely to be some place between the lines. But even if both lines have dip down from somewhere near their intersection, there still is not necessarily a closed anticline.

The main way anticline is detected and checked is by contouring. A closed contour on a map, that is, a contour that curves around to meet itself, indicates a closure, either a closed high or a closed low. Low closed features aren't prospective, so they don't attract much attention. When people talk about closures, they are usually referring to high ones.

The vertical closure (or just the closure) of an oil trap is the vertical measurement of the height, in either time or depth, that could trap any available oil. On a contoured map, you can measure the closure by subtracting the time or depth of the highest point on the structure from that of the lowest point on the high side, the point at which oil would spill out, floating up along the formation.

If you contour the data from two lines, you may see that one of the quadrants between the lines could well be higher than any point on the lines. This is particularly true if the structure is low relief, without much dip. You will probably need more than the two lines to determine that closure of the structure in all directions is likely. The test is in contouring. If, with smooth reasonable-looking contours, the feature can be contoured as not closed, more control is needed. But if it takes a funny-looking skinny strip leading out between lines to keep the structure from being closed, then it probably is closed. Better yet is to have it definitely closed, with lines all around the structure, on which the horizon is lower than on the anticline.

Seismic interpretation of an anticline can be extremely simple when the reflections are good, when the anticline is steep enough to be easily seen on a section but not so steep that it is difficult to recognize, and where there are lines that have lower data on the same horizons in all directions.

You can see from the qualifications in that statement that there are factors that can make anticlines difficult to map. The reflection quality can be so poor that the section can't be picked even in this fundamental way. This situation, though, is, and in general always has been, rare. Whatever the stage of development of seismic techniques, people don't invest a lot of money in shooting areas which, at that stage, don't yield interpretable data.

An anticline can have so little dip that it is not noticeable on a section. This can be true of a small structure, which therefore doesn't have much total uplift or much volume to contain hydrocarbons. Or the structure can cover a large area and therefore have considerable relief. Even with the low angle of dip, it can have room for a large accumulation of oil or gas. Having so little dip may make it difficult to see on a section. One solution to this problem is just normal timing and mapping of reflections. The numbers and the contours show that there is a structure, even if it wasn't noticed on the sections. It can be made more visible on a section by making a section that is compressed horizontally, with the vertical dimension unchanged. The subtle relief becomes less subtle, more easily seen.

An anticline with very steep dip can be difficult to interpret, too. Because sound must be reflected perpendicularly to return to its source, steeply dipping reflections can be recorded only from some distance away from the feature. If the lines do not reach that far, the steep dips won't be recorded at all, so you won't know the structure exists. And if the sections are not migrated, the feature will look broader than it is and maybe not well defined.

Work Exercise 8–1, contouring an anticlinal structure.

PALEOSTRUCTURE

Most deposits are laid down fairly flat. They are mostly deposited on sea beds at some distance from land, so that particles of sediment drift down, forming more or less flat layers. Then tectonics distorts the layers, bending them, breaking them. On a seismic section you see the beds in their current condition—once-flat layers that are now bent and broken, sometimes eroded after the distortion and redeposited. If you take a seismic section, all warped-looking as it is, mark one horizon, and push traces up or down until that horizon is flat, it looks about like it did when it was deposited. Then you can see the approximate paleostructure below it, the reflections below it showing the shape of the deeper formations as of the time the flattened horizon was deposited. This does not apply to a bed that was laid down on a near-shore sloping sea bottom, so you should generally avoid flattening such a horizon.

On the new section, the reflections above the flattened one are not so geologically meaningful. They didn't yet exist at the time the flattened horizon was laid down. They are not needed on the flattened section. If you flattened it manually, the upper part can be removed with scissors. If the flattening was done in a data-processing center, that part could be omitted, not played out onto the film section.

After flattening, you have a section that fairly well shows the structure of an earlier geological time. Other copies of the section can be made with other reflections flattened to show what the structure was when other formations were deposited. The flattened sections can be displayed as a sequence of geological times, allowing a study to be made of the geological history of the area. Studying the geological history is useful to geologists in getting an idea of the forces that acted in the area, and may lead them to develop new prospects.

Another purpose for knowing the paleostructure is that it, rather than the present structure, may be the primary key to finding oil in the area. Consider an anticlinal structure that existed in some long-ago geological time. If it contained a porous formation, that formation probably had water in it. The oil in that water would slowly float to the top of the water and be trapped in the structure. Then, if minerals in the water precipitated out of solution, there would be cementation, possibly enough to make the previously porous formation nonporous. But the trap would have oil in it, not water. The oil would be trapped, not moving. Most substances do not dissolve readily in oil as they do in water and so wouldn't be available to precipitate. Cementation would not take place there. If further tectonic activity occurred after that, the oil would stay where it was, in the only remaining porous part of the formation, sealed in by the cemented rock that had held water. No matter which way the oil trap was tilted or folded, the oil would no longer be free to move to the top of the new configuration. It would stay where it had been before the cementation and tectonics.

The place to look for oil in that area is in the paleostructures, not in the present-day structures. You have to determine not what the structure is, but what it was.

But remember that there were some "ifs" in that discussion. The oil is not always in the paleostructures. If there did not happen to be accumulation followed by cementation, then the oil either was not trapped before the later tectonics or had been trapped and was then shifted to the new high. You have to know the area or make some guesses about it or drill some present structures and some paleostructures. Flattening lets you see where the paleostructures are, so you can then think about what to do. Often, of course, the flattening shows the old structures to be at the same place as the present structures, so you then know you don't have this particular problem to worry about.

FAULT INTERPRETATION

Faults are often critical to the accumulation of oil. They can be critical in either a positive or a negative way. A fault may form a seal by cutting off a structural or stratigraphic feature so the oil is trapped against the fault. The fault may itself do the sealing if there has been cementation of rubble within the fault plane. Or if the fault is a clean break, a seal may exist only if the fault happened to move a shale or other sealing lithology against the updip broken end of a porous layer. On the other hand, if the fault plane contains rubble that has not been cemented, it may form a conduit for fluids. This may allow the hydrocarbons to drift up the fault plane into the feature and be trapped in it or to escape from the feature by drifting up the fault plane out of it. Seismic data doesn't tell you how faults behave in this respect. Wells drilled on faulted features in the area at least tell you how those particular faults behaved. It may be reasonable to assume that other faults in the area will do similarly. As in so many other things about seismic exploration, knowledge of the area is necessary.

Faults are of different types in terms of which way the parts of the broken formations moved. In a normal fault, there has usually been some tensional force tending to pull the rocks apart. An intrusion may have bent the rock up over a greater length than it had occupied. So the rock breaks, and one part drops down lower than the other. Normal faults are the kind most often encountered in oil exploration. The discussion to follow about ways to detect and interpret faults will be mostly directed at normal faults but largely applies to other types as well.

A reverse, or thrust, fault occurs when some compressing force is pushing the rocks together from the sides. The rocks break, and some slide up over the others. The fault tends to be low angle, and the fault plane can curve steeply upward near the surface, in a sled-runner fault. The lower part may even be horizontal, with one formation sliding over another for some distance.

In a lateral, or strike-slip, fault, the movement is horizontal, with one part moving sideways along the break. This type of fault is harder to detect from seismic data than the other two types, as the formations may still be at the same level where they meet on a section. If there is structure, so that a high is moved to a position adjacent to a low, then the fault will look like a normal fault on the section. A strike-slip fault through a series of highs and lows looks like a normal fault with its direction of throw changing, so one side appears upthrown, then downthrown. . . .

A fault in the subsurface usually breaks a number of horizons, not just one. It can be interpreted better, even in interpreting just one horizon, if the breaks in other horizons are taken into account. A fault normally crosses the section at an angle. The best fault evidence is a clearcut break in each of a number of horizons, the line of breaks slanting down across the section. The difference in reflection times between the two parts of a broken reflection indicates the throw of the fault in seismic time, at the level of that reflection.

Another type of fault indication is a diffraction, the down-curving hyperbola-shape at the broken end of a formation or at some other abrupt change in the subsurface. The diffractions at a fault may appear on a slanting line, like the umbrellas of a line of people on a hillside in the rain.

Sometimes faults aren't easy to see, with no clear breaks and no diffractions, so they must be found from other clues. If there is a sharp change of dip in several reflections and that change slants across the section, then it may be along a fault plane. Or it may indicate only a hinge line, with no formations actually broken. There can be just a slanting line of poor reflections or a slanting line of gaps in the reflections. Notice that word "slanting" in each of the types of indication. A fault could possibly be nearly horizontal or nearly vertical, but other explanations of horizontal or vertical alignments on the section are more likely. A horizontal line of diffractions, for instance, is probably from an irregular surface of a formation. Almost any vertical line of anomalous data is likely to be caused by a near-surface or processing problem.

There may be a slanting line of, not breaks, but bends, in the reflections on a section, with the dip remaining the same and the reflections offset. The bends line up like a fault but aren't clearcut breaks. They look like the rocks bent sharply but didn't quite break. Seismic reflections aren't perfect images of the subsurface, though. Diffractions could round the sharp edges, making a break look like a bend. Although a migrated section should be fairly free of the diffractions, even it isn't a complete picture of the rocks. This is a matter of judgment. Many, probably most, people draw a fault in such a situation. I prefer not to, but to generally decide a phenomenon is not a fault if it isn't pretty clearly a fault.

The angle of the fault as it appears on the section is not a true angle, for two reasons. It is made up of a combination of vertical time and horizontal distance. And unless the direction of the line is exactly perpendicular to the fault trace, the angle on the section represents only a component of the angle of the fault. The component, not the true fault angle, could be determined, if desired for some reason, by converting the broken ends of the reflection to depth. On another piece of paper, these depths can be plotted against horizontal distance read from the shot point numbers on the section. The angle can then be measured on the plot. The trace of the fault is obtained by plotting on a map the points at which the fault cuts one horizon on several lines. The true angle of the fault could then be calculated from the relationship between directions of line, fault trace, and component. This is not normally done, though. Seismic fault evidence is not exact and dependable enough to warrant it, and the angle of a fault is not particularly useful in the search for oil.

If a seismic line runs exactly along a fault, then the fault should appear horizontal on the section. That's right, except that it would be very hard to detect the horizontal fault on the section. What really happens is that for practical purposes we can detect only faults that appear on the sections to have angles of something like 30 or more degrees. The low components of fault angle are pretty much lost to seismic exploration.

In many situations the throw of a fault cannot be determined with any reliability. If the fault is in a zone with many reflections of about the same quality and appearance, for instance a sand-shale sequence, it may be difficult to tell one reflection from another. If reflections on opposite sides of the fault happen to meet, it may appear that there isn't a fault there at all, that reflections just continue across the spot. If the reflections don't happen to meet, the fault may appear as a break in the reflections, but with very little throw. It is natural for the eye, in trying to correlate across the fault, to look for similarity between a reflection and another one nearby, rather than a farther one. Cutting the section apart along the fault with scissors and then shifting the two parts and looking for the best fit may help.

Correlating across a fault is easier if the parts of the geologic section on the two sides happen to be equal in thickness. They are equal if the fault occurred after the sediments forming those parts of the section were all deposited. But if the fault is a growth fault, it occurred in stages and deposition went on between the times of movement. The downthrown side was in deeper water, or the upthrown side was above the water. The downthrown side received more sediment, or was eroded less, or both, than the upthrown side. The difference in weight of overburden may even have caused some of the fault movement. The downthrown side, with more sediment, became composed of thicker rock units. On a seismic section it looks like the upthrown side, but with the reflections farther apart. It is harder to recognize a similarity between the two sides than it would be if the reflections were the same distances apart. Again, cutting the section apart and shifting the parts may aid the correlation. If may even enable you to determine just when, in terms of the reflections, the movements occurred. The faulting was probably intermittent, not continuous. With good records, you may be able to say that between a certain two reflections the fault was a stationary scarp, then there was movement, etc.

If all the fault movement occurred before the sediments were deposited, then it will appear like only a part of the growth fault. The fault would have been a scarp, with sediments collecting at its base. Such an early fault, if it has a large throw, appears on a seismic section as a large break in a reflection, with sometimes a talus pile visible as a nearly reflectionless zone on the downthrown side. But reflections above it are not broken. They either continue straight across or may be somewhat bent by differential compaction. That is, some of the thick sediments on the downthrown side will have more total compaction than the thinner layers of the same sediments on the upthrown side.

Faults can have effects on the velocities of rocks. These effects are likely to cause horizons below the fault to be misinterpreted. A reverse fault pushes a normal thickness of a formation up over another similar thickness of the same formation. If that is a high-velocity layer, then the sound traveling down through two layers of it arrives at the surface sooner than the sound going through just one layer. So reflections below the overlap appear higher, like an anticline. This is particularly difficult to interpret, as it looks so geologically reasonable, as though a real anticline had caused the fault at that point. However, if the double layer is low-velocity material, it could cause the opposite effect. The reflections below it would appear low, though, and wouldn't cause people to drill wells. A normal fault also can have velocity effects where a formation is missing at the fault. In general, be wary of anticlinal features that happen to be right under faults.

Another severe problem was discussed earlier under diffractions. It is the risk of misinterpreting a segment of a reflection that has a fault and accompanying diffraction at its updip end, as an anticline.

The worst problem in interpreting faults, though, is in mapping, drawing a fault trace from line to line. Some faults with large throw, or in an area without many faults, can be connected reliably. But these are a very small minority of the faults you will see on seismic sections. Most faults have little, or indeterminate, throw. Most are found in areas that have many faults. You interpret the faults on the sections. Then comes time to map.

Let's say that one loop has some fault indications on two parallel lines. On one line are four down-to-the-north faults: 1, 2, 3, 4. On the other are three, also down-to-the-north: A, B, C. How do you connect them? Is 1 the same fault as A, then 2–B, 3–C, with 4 a short one that doesn't reach the other line? Or maybe 3 is the short fault, or 1. Maybe A connects with 4, and all the rest are short parallel fault slivers. Maybe 3 and 4 join together to meet the other line at C.

The only clue you have from seismic data is correlation, this time correlation not of reflections but of the fault indications. If 2 and 3 are very close together and so are B and C, but the others are farther apart, that is a clue that 2 may be the same fault as B and 3 the same as C. Or if 2 and A have more throw than the others, maybe those two should be connected. Or if 1 and B look alike, maybe by each having a similar adjustment fault, that may indicate that they should be connected.

A loop may make it easier to see how faults connect. But if the faults are clear on two parallel lines, that probably means that they cross those lines at a large angle. So they may be nearly parallel to the intersecting lines and may be very hard to interpret on those lines.

In Exercise 8–2, correlate the faults on two parallel lines.

Usually one fault indication looks about like another, so you can only guess which ones to connect. In that situation, it is sensible to connect nearby ones and to use some geological hints to help you decide what direction the trend of the faults is likely to be.

But—**warning!** If you do have fairly good seismic clues, be careful about distorting them to fit a preconceived idea of what the geology is supposed to be. Maybe the preconception is wrong. Or maybe it is regionally correct, but your particular area is unusual. You are doing the interpretation to discover things that weren't known, so discover them, don't just confirm an earlier guess.

On the other hand, real geologic data is data and should be combined with your seismic information to get a total picture that neither could achieve alone. Working with a geologist, so you apply both geological and geophysical training, the two of you can probably work things out better than either one of you could.

In areas known to have many faults, there is pressure on geophysicists to map the faults. They have tried. The efforts were often so poor as to be comical. But the interpretations were put on maps and drafted neatly. The maps looked professional—wells were drilled, based on those maps! Granted, many faults are clear enough on the sections to be mapped well, but there are also highly faulted areas in which the faults are very difficult to map seismically.

This problem of connecting faults on a map is a devastating interpreting problem. Faults are an important part of the mapping; in some fault-trap situations they are the most important part for finding oil. Connecting the wrong fault indications must many times have caused wells to be drilled in wrong places, not only drilling dry holes but missing oil fields that might otherwise have been found. You're not to blame; you're trying to get 3-D information from 2-D data, and the clues just aren't in it. If you start to feel sorry for yourself, think of the poor geologist, trying to get 3-D results from 1-D wells.

Work Exercise 8–3, to see what information you can get from mapping faults.

Suppose you do map some faults and feel you have done it fairly well. How can you find oil with them? If you have contoured a plunging anticline, or a nose, and a fault cuts across it, then the fault, if it is a seal, may form a trap. Notice that it often isn't important which side is downthrown, the cutting off is what counts. Or a faulted anticline may have oil trapped downdip from a fault and not on the crest—or the other way around. Even a curved fault trace on a monocline could trap oil. In some reef areas, the faulted reefs may be more likely to be productive than other reefs because the oil can migrate up the fault planes into the faulted reefs.

While we're on the subject, try Exercise 8–4, which is like the preceding exercise but with a different fault pattern.

One of the most difficult fault mapping situations is one in which there are a great many faults, closely spaced. Then there is little to indicate which fault crossing should be connected to which other one. Even the grain of the faulting may not be determinable. You could interpret the area as covered with many faults running east-west and then wonder if you should have connected them all north-south.

Another problem is that there are occasions in which a fault is clearly apparent on two lines that are not far apart, but cannot be seen at all on a line between them. What can you do—force a fault on the section between, throw out one or both of the faults, let two faults approach the middle line but not reach it, draw them at strange angles so they miss that line without ending so abruptly—who knows? It's a matter of personal judgment in each case.

One way to detect a horizontal-movement fault is by correlating parts of a contoured map of a horizon. Sometimes you can cut a map apart along a line of change in the contours and slide one part to a better fit with the other. Look for a better match of the highs and lows across the cut. If you do find a good fit, the line of difference may be a strike-slip fault, with the lateral movement indicated by the distance you shifted the parts of the map.

On the subject of mapping faults seismically, let's digress a bit to point out what could be done in the field to alleviate the problem. The best solution is a 3-D survey, with lines so close together they cover the area with control. With that density of information, you can trace a fault's progress from line to adjacent line, all across the area. Or the 3-D survey can give you a time slice, which is a horizontal rather than vertical section. On it you can see the fault cutting across the reflections. Wonderful! But terribly expensive. Your organization's budget will stand for it only in very special cases, normally only for already-discovered fields.

There is another, cheaper, solution. As you might expect, it is not as good. Shoot lines with some width to them or shoot conventional 2-D lines, but in adjacent pairs rather than in lonely isolation. With either of these techniques, a fault crossing will show, not just that a fault crossed the line or lines, but also its map direction. You see where it was headed when last seen. You can more reliably connect two fault indications that are pointing at each other. A caution with the two-line method—the surveying must be ultraprecise, just as with 3-D shooting. The fault directions will all be misangled if one line is plotted a little farther ahead than the other.

Now work Exercise 8–5, contouring such pairs of lines.

SALT DOMES

A salt dome is a tall column of salt that has pushed up through other formations from a layer of salt. The salt moves upward by gravity, as the rocks overlying the salt layer are heavier than salt. The salt is more plastic, more easily deformed, than the other rock, so it can flow slowly. The salt column is like a liquid working its way up through a heavier solid.

Most salt domes encountered in oil exploration have not reached the surface of the earth. The nearer the salt approaches the surface, the less weight of overlying rock there is, so the likelihood of its breaking through is diminished. Also, salt nearing the surface may encounter fresh groundwater and be dissolved and carried away. A salt dome is likely to have a layer of cap rock immediately above it that was produced by this leaching away of the salt. Groundwater dissolves salt, leaving only the insoluble parts that were in the salt. More salt rises from below and is dissolved, so a cap rock may be quite thick, say some hundreds of meters. It is made up largely of anhydrites and may contain commercial quantities of sulfur.

The salt movement not only bends the overlying rock but also breaks it, so there are usually many faults associated with a salt dome. As the salt moves upward, the formations above it are bent upward in a curve that is longer than the rock layer was before bending. Rock doesn't stretch very much, so it breaks in a characteristic pattern. In a cross section through the dome, this pattern is composed of faults angling out above the dome, downthrown toward the center, the overlying layers dropping down in stages. The salt layer that produces the salt dome is often very deep, probably because a shallow layer of salt wouldn't have enough overburden to force it upward. On a seismic section, a salt dome is often seen as a column that extends downward to the bottom of the section.

The salt in the dome is a mass of pretty much the same velocity, so there are not many reflections in it. The sides of the dome are very steep, often vertical. Sometimes the dome bulges outward. The upper part may even be a separate inverted teardrop shape. All this steepness means that reflections from the flanks of the dome may be very difficult to record seismically. A near-vertical surface could not be recorded by the nearby part of a seismic line but only by a part far away from the feature. The energy that struck it perpendicularly would be traveling nearly parallel with the surface of the ground, so it would reach the surface at some distance from the reflection point.

The dome will be seen largely as an absence of reflections. The gentler dips of the top of the dome can be recorded from nearby and above it so that part of the interface between salt and other rock may show up well on a seismic section. Migration of the sections gives an improved view of the shape of the salt dome.

Work Exercise 8–6, picking salt outlines.

If you see a salt dome on sections and map it, the next step is to find prospects over and around it. This is difficult, with the many possible traps and the difficulty in mapping the fault pattern correctly. Oil and gas are found associated with salt domes, not in the dome, but in the surrounding rock in traps formed by the distortion of rock caused by the upward movement of the salt. A layer may be bent upward and broken off at the salt, so the oil moves up into the upcurved part. The salt is not porous, so it forms a seal. Oil may be above the salt in a layer that has been pushed upward by the salt movement, forming a local anticline above the dome. Oil may be found in traps against some of the many faults associated with the salt movement. The oil may be on one side or another of the dome.

So discovering a salt dome does not put you very far along the way to finding oil. For the same reason, a salt dome that has had dry holes drilled on it is not condemned as to potential for oil. As a geologist said one time when we were considering whether to lease a dome with five dry holes on it, "That's the first five we won't have to drill."

In some salt dome areas, particularly in the Gulf of Mexico, bright spots and flat spots are helpful in finding which apparent traps are likely to contain gas. A bright spot is a locally strong reflection that may be caused by the different velocity contrast between rock layers because of local gas content. A flat spot is a locally flat reflection from the flat surface of a liquid in the subsurface. They will be discussed in general application later in this chapter.

OTHER DIAPIRS

A diapir is any rock mass that has been forced upward in a fluid manner into overlying rocks. Salt domes are diapirs. So are shale masses that were squeezed upward. And so are igneous intrusions that flowed upward into other rock.

Shale diapirs are like salt domes. On seismic sections it is difficult to impossible to distinguish between the two. But the two can also trap oil in similar ways, so in some areas the difference may not matter.

Igneous diapiric intrusions may also appear like salt domes on the sections. But they are not as likely to be associated with oil. The heat of the igneous flow melts and metamorphoses the rock near it, destroying any oil in that rock and also the porosity of the rock.

OTHER SALT FEATURES

There are other shapes that salt can take that can be associated with oil accumulation. In the early stages of diapiric movement, before a dome has formed, or if there is not enough salt to form a dome, the bulge upward of lighter salt under heavier rocks can form a salt swell or a salt pillow. The rocks above are not so likely to be broken in these lesser movements but may be bent enough to form anticlines that can have oil trapped in them.

In a totally different way, by solution, salt can cause features to form that can trap oil. In one area, the following explanation was proposed to account for some small oil fields. Water got into a layer of salt in the subsurface and dissolved the salt in spots. In a spot where the salt was removed, the overlying rock sank down into the hole, filling it. Later sedimentation filled in the low place in the surface, depositing a layer that was flat on top, thus thicker in that spot. Still later, water removed the rest of the salt in the salt layer. The overlying rock settled down. But one layer was thicker where the original solution had taken place, so the layers above that one were bent upward over the thick spot. Oil could be trapped in the anticline. These anticlines are small but numerous and close together.

On seismic sections these features can be seen as slight uplifts of some reflections. The horizons below them are not uplifted, and above them the uplift is progressively less for shallower formations. So the feature can be seen as a bend in a reflection between flatter reflections above and below it. Flattening a horizon—almost any horizon—helps to make the features visible.

REEFS

For seismic purposes, a reef can be thought of as any piled-up-looking mass of limestone. There are more subtleties in a geologist's definition, more types of piles of lime, but the distinctions are generally beyond the seismic section's discrimination. On a section, we can't tell if a feature is reef in place or a heap of reef detritus. Fortunately, the distinctions don't matter as much as just finding the pileup. Either type can be porous and full of oil. We can't be as discriminating as a geologist, but we're more so than a seaman, who calls any shallow place in the sea a reef. I'll use this geophysical definition, the only practical one when we are considering features on present-day seismic sections.

Reefs can be clear and easy to recognize on sections, or less clear and more difficult to detect, or totally invisible. Unfortunately, the clearest often are not the most desirable.

A textbook reef can have a strong reflection at its top and another at its base, so its shape is clearly defined on the section. Reefs develop on limestone layers, so there is likely to be a reflection from the top of the layer that then humps up over the reef and maybe appears to split, one part going over and one under the reef. These excellent reef reflections tend to occur in shallow parts of the section, where the limestone velocity differs radically from the velocity of shale that has not been compressed very much by overburden. Because of compression, the velocity of sound in shale increases with depth, so a deeply buried reef might be surrounded by shale that has a velocity about the same as the limestone.

Pick the reflections around a reef in Exercise 8–7.

A more difficult case is the reef that appears on the section as a weak reflection going over the reef, with probably no reflection at the base. This type can be difficult to map. Lines crossing the feature may even disagree, some showing the feature and some not. An especially weak reflection from the upper surface of the reef can be caused by gas in the reef. The low-velocity gas reduces the overall velocity of the limestone mass. If this reduced velocity is about the same as that of the surrounding rocks, then there may not be a reflection at all from the top of the reef. A reef that is hard to see may well be a productive reef. It may produce gas, and below the gas there may be oil. Of course, the reflection may be poor for other reasons, so just being hard to see doesn't necessarily mean that a reef has a lot of gas, or oil and gas, in it.

The top or base of the reef may not appear on the section at all, so the reef can be found only from other, more indirect clues. I'll lump them and the direct ones together in a list of kinds of seismic evidence for reefs.

Reflection from reef top. This is the most direct evidence. If the reflection is a good one, it allows the shape of the reef to be mapped.

Reflection from base of reef. This helps to show the shape of the reef, when the top can be seen, but alone is not a good clue to the reef.

Shape. A reef grows upward on another surface. This gives it its characteristic shape on a seismic section, of a lump on a normal-looking reflection. The lump may be flat-topped, from the reef's growing to near the flat surface of the water. It may be a rounded mound, all that was left after some erosion of the flat-topped reef. It may be rounded because it is not the original reef where it grew, but a pile of reef detritus.

A reef may appear on a seismic section as two mounds with a somewhat lower part between them. This indicates a reef with a lagoon within it. Wave action can't reach into the lagoon very vigorously, so the lagoon is likely to be muddy. The mud fills pore spaces, leaving less room for oil to collect in than the higher parts of the reef.

Dead zone in reef interior. This is an absence of the reflections normally encountered at that level, the reflections from bedding around the reef. It is especially noticeable when a strong reflection at that level is locally absent.

Reflections from within the reef. These may bend like the layers in a compost heap, showing layers formed as the reef built up, or as reef detritus piled up, in stages. Their curving shape contrasts with the flatter reflections around the reef.

Velocity effects on deeper horizons. These are usually in the form of velocity pullup. The reef is a thick chunk of high-velocity limestone, often surrounded by lower-velocity sands and shales. Sound gets through the lime faster than it goes through the other rocks, so reflections from deeper horizons under the reef arrive at the surface quicker than those that go through the slower materials around the reef. On the section this causes the reflections below the reef to be bent upward under the reef, pullup. The effect is greatest just below the reef, where the sound goes through the reef. Deeper, the combining of near and far traces in the CDP stack includes energy that undershot the reef, going at an angle and passing to a reflecting point below the reef and up at an angle to the surface without going through the reef, so the pullup is less. In some geological situations, the reef velocity is slower than the velocities around it, so instead of pullup, there is pulldown, with the horizons just below the reef appearing to sag rather than be uplifted. And there is a borderline case in which the velocity of sound in the reef is about the same as in the surrounding rocks, so there is neither pullup nor pulldown.

Drape above the reef. Limestone is not nearly as compressible as shale. So differential compaction can cause the layers above the reef to drape over it. Deposition takes place over a reef, burying it under flat layers. More time passes, the compressible shale is mashed down and the hard limestone is not. That makes overlying layers bend, staying high over the reef and sagging down around it. The beds below a reef are either unaffected or apparently pulled up or down by velocity, while those above are bent up over it. This appearance, when you can see no other evidence of the reef, can in some areas be conclusive evidence ot it. Experience in the area lets you determine the meaning of the drape.

Diffractions. Sometimes the edges of a reef are abrupt enough to have diffractions, an effect of sharp discontinuities in the subsurface, like where beds are broken at faults. The change from flattish top of a reef to steep side is such a discontinuity. So some reefs may be detected by diffractions curving smoothly off their edges.

Geologic setting. Reefs can grow only in sea water and only in places so shallow that the top of the reef can be within a few meters of the top of the water. They don't grow at all above high tide. They grow vertically when continuing subsidence of the land or rising of the sea level makes the water get progressively deeper. Their growth is inhibited if they are in places where mud can collect and prevent the coral from getting its nourishment from the sea. So buried reefs are to be found near the edges of ancient seas, mostly on shelf edges. The water is shallow on the shelf and washed clean of mud at the edge. If there is a sloping reflection that abruptly becomes steeper, then that place of change, that hinge line, is a likely place for reefs to grow. Look all along the hinge line for reefs. Also, farther back on the shelf, any local slightly high places would provide chances for reef growth. Being elevated would keep mud from settling there. There can be a complex of pinnacle reefs on one of these elevated places. These reefs are usually small, on the order of a kilometer across, with many of them closely spaced in a confined area.

In an area where upper surfaces of reefs are not visible on seismic sections and where the reefs must be detected by drape of other horizons over them, the drape is usually slight and therefore difficult to see if the horizons do not happen to be fairly flat. It there has been folding after the reef growth, then it may be helpful to flatten some reflection in order to see the drape without the added complication of the folding. For this purpose, it may not matter much what horizon is flattened as long as there isn't much difference in depositional or tectonic attitude between it and the drape you want to see.

It is probably best to select a prominent, good quality reflection to flatten. At least, it can be picked more correctly than a poor one. The reflection can be either above or below the draped horizon or can even be that horizon itself. In these three situations, different configurations will indicate the drape. Flattening a horizon above the drape will cause the drape to appear with less relief than it actually has. The closer the flattened horizon is to the drape, the less drape will be apparent on the new section. In this case it is best to flatten a shallow horizon, the shallowest good one on the section. If a horizon below the reef is flattened, the drape should appear with its correct relief except for the effect of velocity pullup of pulldown. If the draped horizon itself is flattened, as might be done if it is the only reliably continuous reflection on the section, then horizons above it will show apparent sag over it. Those below should also show a sag, modified by velocity effects.

There is a problem in shooting an area of pinnacle reefs. The close spacing and small size of the reefs make it necessary to shoot many closely spaced lines in order to detect them all. But when you are interpreting such an area, there may not be a close grid of lines, for any of several reasons. The group of reefs may have been discovered by a widely spaced grid of reconnaissance lines, so when you are interpreting those lines, you don't yet have detail control over the area. And until and unless some of the reefs have been drilled and found to be productive, your company may not be interested in spending the money to shoot a close grid. Even after that, there is a period in the development of the area during which the special needs of such frequent, close, small prospects are not recognized by management. For most of the time you work such an area, the control is likely to be inadequate for it.

The result of this is that you will have lines that cross the tops of reefs and other lines that just clip the edges of reefs. There will be no way to tell whether a reef that appears tiny on a section is actually tiny or just was not shot across its central part. At first, when only reconnaissance lines have been shot, there may be only a few reef indications on the sections.

OTHER TRAPS

We've covered some of the ways oil can be trapped, but not all. Ingenious people are constantly thinking of more situations in which oil may be found and also are working out explanations of the trapping mechanisms of already-discovered oil. Pileups of sand can be oil prospects. Their changing thicknesses can be handled in the same way as reefs, either from a known example in the area or calculated from a known or estimated sand velocity for the horizon. The salt-solution trap described is an example of the subtlety. The crust of the earth is complex. It is old, and many things have happened to it. There can be many subtle ways in which some porous zone can be high and sealed with nonporous material. Oil has been found in the rims of buried meteorite craters and in granite wash. It is often found by accident. The accidental discoveries are a clue to how inadequate are our knowledge of the earth and our techniques for learning about it.

BASEMENT

The word "basement" applies to some very different things. Its basic meaning is the top of the igneous rock under the bottom sediments. In this sense it is probably the point below which there is no expectation of finding oil. An extension of the word is its use to mean economic basement, which can be either the igneous basement or some layer above it, perhaps a metamorphosed zone or merely the deepest formation expected to have oil. Then there is the research geologists' acoustic basement, the deepest somewhat continuous reflection on the section.

Whatever kind of basement you are concerned with has special problems. It is deeper than your main horizon. The field and processing operations are usually designed to produce good data at the level of the main horizon. And the deeper information is likely to be poorer because seismic energy must travel through more rock layers to and from it. So you will often be trying to map a very poor reflection that is discontinuous. It will probably have more dip than the shallower layers. It may have more abrupt features, for instance faults, where the shallower formations are more gently draped over it. When it is igneous basement, it may have an irregular surface that, on unmigrated sections, produces a "horizon" composed of many diffractions.

STRUCTURE AND STRATIGRAPHY

There is much talk of the two main uses for seismic sections—to determine structure and stratigraphy. But those words are borrowed from geology and don't quite apply to geophysics as they are used.

The conventional use of seismic sections is said to be to determine structure. But what it really determines is the configuration of reflections, or structure-like shapes. In conventional interpretation, structure, the result of tectonics, is found and also the similar shapes of reefs, erosional features, drape over differential compaction, meteorite craters, sand dunes.

Part of stratigraphic interpretation is accomplished by determining from a section parameters other than the alignments of reflected energy—parameters like velocity, amplitude, frequency, phase. These parameters yield information on lithology and content of pore spaces. The study of these parameters has come to be called seismic lithology.

Then another part of stratigraphic interpretation, the part called seismic stratigraphy, uses the structure-like shapes but in a more general way. Instead of the configuration of one reflection, it is concerned more with packets of reflections, as to how nearly parallel to one another they are, how strong the reflections are, the shape of the packet as a whole. It gives information on the depositional history of the geological section.

So the types of seismic interpretation of sections tell us things about (1) the configurations of horizons, (2) the constituents of the rocks, and (3) the deposition of the rocks. These distinctions are based on the different ways people work with seismic data. Separating interpretation into the two conventional types, structural and stratigraphic, is in this respect artificial. Probably each of the two conventional types can be found to some extent in each of the three I suggested above. We've been considering the first of the three, configuration of horizons. Now we can take up the other two.

CONSTITUENTS OF ROCKS

Lithology is not directly determined from seismic data, but there are characteristics of the data that can be used to provide some evidence about the kinds of rocks and their fluid contents. Measurements made on the seismic data yield information on velocity of sound, frequency of the reflected energy, polarity of that energy, and the amplitude of the reflections.

AMPLITUDE

A seismic reflection is strong or weak depending on the difference in velocities between the rock layer above the reflection and the one below it; the greater the difference, the stronger the reflection. This is complicated somewhat by reinforcement and cancellation between nearby reflections, but it is a fairly good general principle. A strong reflection is one in which the trace swings far to the side. The strength, or distance the trace swings, is the amplitude of the reflection. When lithology or fluid content of one of the rock layers varies, the velocity contrast is likely to vary, causing a difference in the reflection amplitude. A large increase in amplitude over a short horizontal distance is a bright spot. A decrease in amplitude over a short distance is a dim spot.

Amplitude is useful in seismic interpretation in several ways. The strength of a reflection is an aid to correlation. It helps an interpreter distinguish one reflection from another. You may recognize it by its being stronger than other reflections above and below it or, sometimes, by its being weaker.

A bright spot indicates a local difference in velocity of a rock layer, increasing the contrast in that place above the usual contrast for that horizon. If locally the layer is porous, or gas-filled, or composed of coal, the velocity of sound will be less than normal for the layer. If that layer was already the slower one of the two adjacent layers, then there will be more contrast and therefore a higher amplitude. And even if it was the faster layer, the change may be so great as to make the contrast the other way around—and also greater. In sand and shale, a bright spot can often mean that there is gas in a sandstone. Or the bright spot may exist because of some lithologic variation. And even if it does indicate gas in the formation, the amount of gas is not necessarily large enough to be commercial.

A good thing about detecting gas, even if it may not be in commercial quantities, is that there may also be oil. The velocity of sound in oil is slower than in water, the usual fluid in the rocks, but not enough slower to create much of a bright spot. So you will not find oil directly by finding a bright spot. But where there is gas, there may be oil below it.

Limestone is frequently higher in velocity than the overlying rocks, often a lot higher, so there is a strong reflection at the top of the limestone. Gas in the top of a limestone body, for instance in the top of a reef, makes that part of the limestone slower and therefore nearer to the velocity of the rocks above it, so the reflection is weaker, a dim spot. Again, if there is gas there may be oil. For this reason very productive reefs sometimes have weak reflections at the top and so are very hard to find. A bright spot is shown in Illustration 8–1.

FLAT SPOT

A flat spot is a different phenomenon, not an amplitude anomaly. It is a reflection from the flat upper surface of a liquid, that is, a contact between a liquid and a gas. It is recognizable only when the other reflections around it are not flat, so its flatness stands out. When there is dip in most reflections but a flat reflection exists over a limited area, it may be the upper surface of a liquid. It could possibly be that the flat reflection is a multiple of some strong flat event somewhere shallower in the section, but this isn't likely unless the shallow layers are flat or there is another top-of-liquid flat spot at that shallower depth. There is a flat spot in Illustration 8–2.

I said at the start of this topic that a flat spot was not an amplitude anomaly. It isn't, but the same segment of reflection can also exhibit an amplitude anomaly so a spot can be flat and also bright. There is a strong velocity contrast at a gas-liquid interface, so the reflection is likely to be bright. It may also appear to be a polarity anomaly. If the flat spot happens to line up with a reflection, the velocity differential may be in the opposite direction from that of the reflection. So you can see a reflection that looks flat, bright, and of reversed polarity.

Normally a flat spot indicates the contrast between a gas and a liquid, not between two liquids. The difference in velocity between water and oil isn't great enough to show up clearly on a section. But it can be an indication of the base of the oil for a different reason. If, after the oil was trapped, minerals moving through the water went out of solution and cemented the pore spaces, those same water-soluble minerals would not be likely to be dissolved in, and precipitate from, the oil. So the rock below the oil would have a changed, less porous lithology, while the part where the oil was would remain porous. This difference in lithology could appear as a flat spot on a section.

There are two situations in which a "flat" spot can be tilted. First, the flat spot just described would still exist if the formation was later

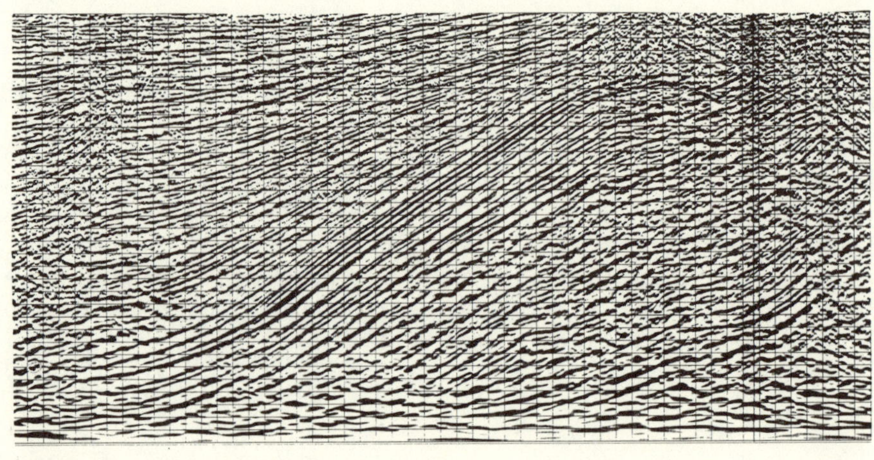

a

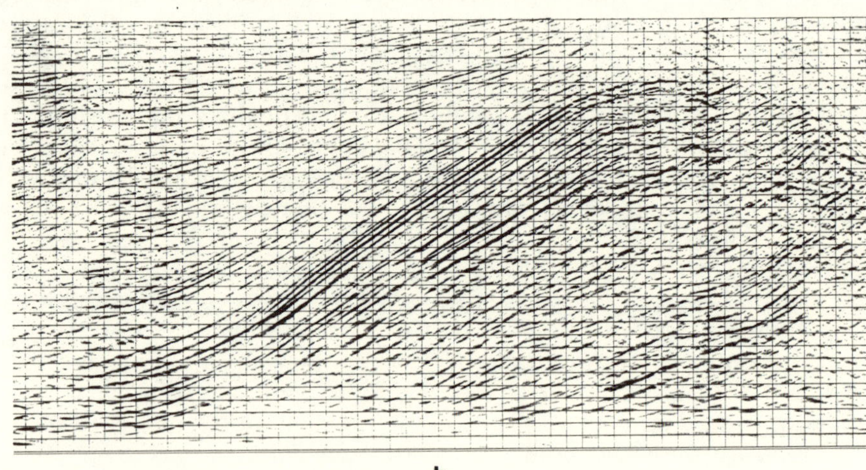

b

Illustration 8–1 Bright spot (a) on conventional section and
(b) on true amplitude section
(courtesy Digicon Geophysical Corp.)

tilted. And in that case, it might retain the oil in the tilted trap. The second situation is the case of a flat spot that is flat in the subsurface but appears tilted because of the velocity effect of shallower horizons. In both these situations, the "flat" spot stands out by its differing from the dips of the reflections all around it. In either case, theoretically, if you knew enough about the situation, the section could be corrected or adjusted to make the spot flat. On the section with the spot actually tilted, you could flatten some other horizon to restore the section to a

Illustration 8–2 Flat spot
(courtesy Teledyne Exploration)

time before the rocks were tilted. The section with the apparent tilting caused by velocity could be converted to depth to make the spot flat if you had good enough velocity information available.

DIRECT HYDROCARBON INDICATORS

These anomalies—the bright, flat, and dim spots—are often referred to as direct hydrocarbon indicators, or DHIs. Those terms are not quite correct. What is indicated is either gas or lithologic variation. The gas can be hydrocarbons, but it can as well be any other gas in the subsurface, like carbon dioxide or hydrogen sulfide.

POLARITY

In the discussion of problems on sections, the polarity of a section was described as the direction a trace swings in response to movement of the ground. The same term applies to individual traces. And the polarity of a reflection is the direction the trace swings at the start of the reflection.

The effects of different fluids in the rock and of different lithologies can affect the polarities of reflections. If a layer has a local change in velocity because of gas or a change to coal or something, its velocity relationship to a bed in contact with it can change enough to make the velocity difference be in the other direction. For instance, if a low-

INTERPRETING SEISMIC DATA

velocity rock is over a fast one and the change makes the fast layer even slower than the other, then a reflection swings in the opposite direction. A peak abruptly becomes a trough or the other way around.

A polarity reversal, like a bright spot, is an indication of change in velocity of a layer. It may indicate gas in the formation.

FREQUENCY

The frequency of sound, in cycles per second, can be obtained from a seismic section. It could be calculated by counting the wiggles in one second of reflection time, but that would only be an average for the whole second. A more specific measurement can derive the frequency from the time span of a single cycle, by dividing that time into one second. One cycle is one peak and one trough. This gives the frequency at every cycle. A plot of instantaneous frequency, made by processors, uses a calculation of the frequency continuously down the trace. It is usually displayed in color and is often superimposed over a conventional wiggle trace section. The changes in frequency give some hints at the nature of rocks in the subsurface. This works particularly when the display can be calibrated against several wells in the area.

VELOCITY

Velocity of sound in the subsurface is determined, rather crudely, from the overlapping common-depth-point data. A CDP gather is a collection of the seismic traces that reflected from one point, the traces that are to be combined in stacking into one trace to represent that point on a seismic section. In the gather, the traces differ in horizontal distance between source and receiver. A near trace is from a source and receiver that are both close to the surface position directly above the depth point. A far trace comes from recording with the source and receiver both far from the spot above the depth point. In both cases, the source is the same distance from that surface point as the receiver is. And to keep this discussion simple, I am assuming that the subsurface rock layers are all flat. If they aren't, the reflections are shifted a little from the midpoints, but mostly that is ignored in processing.

In the gather, a certain reflection that is at a certain time on a near trace is at a greater time on a farther trace because the sound traveled a longer distance in slanting to the reflector and back to the surface. On the near trace, the path down and up was more nearly vertical. This difference, with time increasing from trace to trace, makes the reflection curve downward as farther traces are reached. The curvature must be removed to stack the traces so the energy of a reflection on different traces will combine. The degree of curvature depends on the velocity of sound in the rocks the energy passed through. The slower the velocity, the more curvature.

The amount of movement of a trace necessary to remove the curvature is a function of velocity and so can be used to calculate velocity. But deeper reflections have less curvature, so the method is less sensitive to variations in velocity in the deeper parts of the section.

So velocity can be determined from the seismic data itself but not, in that way, very accurately. The accuracy is greatest if the horizon is shallow, where there is a steeper slant to the path of the sound. This makes for a greater curvature of the reflection on the gather, enough curvature to work with effectively. And the velocity determined is average velocity, the velocity of the whole path down to and back up from the reflecting horizon. That velocity does not involve the formation itself, but only the layers above it, all the way to the surface.

To learn things about the composition of the layer, we need its own velocity, the interval velocity of the layer. To get that, we can take the average velocities of the top and bottom of the layer and derive the interval velocity from the difference between them. You have the times to the two reflections, obtained from the section. With the average velocities you can calculate their depths. Get the differences in times and depths by subtracting and then calculate the velocity from one to the other. This is the interval velocity, which may help some in deciding what the lithology of the layer is. The interval velocity can be obtained more accurately if the layer is thick, so there is enough travel time between top and bottom of it to make distinct measurements at the two surfaces. The velocity is most correct if the horizon is shallow and thick. The trouble is that we often want most to use the method for layers that are deep and thin.

Once the velocity of a formation is calculated from the seismic data, it can be used to help decide what kind of rock makes up the formation. It isn't definitive. The velocities for kinds of rock vary widely, depending on porosity, material filling the pores, the mixture of rock in the formation. There might actually be a single definite velocity for, say, a solid block of pure silica. Sandstone isn't a block of silica, though. It is a mass of grains of silica with open spaces between them. The grains vary in size and shape, both of which affect the amount of space between them. The spaces can be filled with water, oil, or gas. There is likely to be a mixture of other things with the sand, like shale or coal. We speak of a shaly sand or a sandy shale.

As you can see, the vagueness is a matter of definition. The word "sandstone" is so broad that it includes a number of varieties of sandstone. Because of all this, words like sandstone, shale, limestone, coal cover rocks that have a broad variety of velocities. The velocity range for one type of rock overlaps others, so a slow limestone and a fast sandstone may have the same velocity. And rocks are compressible, so they have higher velocities when they are buried deeper. Shale is especially compressible. As it is compressed, the fluids are squeezed out of it. The velocity of shale increases at a steady rate as it is buried deeper.

A more detailed analysis of velocity can be made by using the stacking velocity determinations together with calculations made from the strengths of the reflections. The reflection strength depends on the reflectivity of the rock interface, the difference in velocities of the rocks making up the interface. The greater the contrast between velocities of two kinds of rock, one overlying the other, the stronger the reflection between them. If you knew the velocity of one of the rock layers, you could calculate the other. Then having that second one, you could determine the next one from the strength of the next reflection. And so on. These are more precise velocity computations, one relative to another, but they could wander astray in overall correctness. So the stacking velocity is used to keep this mass of relative velocities on track.

That method enables people to determine velocities continuously down a seismic trace. They can then be plotted like a sonic log. This plot is a synthetic sonic log, or inverted trace. Notice that the stacking velocity calculation uses the traces that will be put together to make one stacked trace, and the calculation of reflectivities uses the reflections on that one stacked trace. So this synthetic sonic log is made from what, on the stacked section, is only one trace. Another can be made from the next trace. All the traces of a section can be converted to synthetic sonic logs. This is a different-looking section, one that is more useful than a conventional section for investigating rock types. It isn't absolute, either, but it's considerably better. A particular use of it is in finding the more porous parts of the subsurface, so wells can be drilled in the porous places where there is more room for oil to accumulate.

A section made up of these synthetic sonic logs alone is not very visual, that is, the velocities you need to see do not stand out from each other very well for determining porosities from them. So another trick is often used to make the display more useful. The velocities are machine contoured. The contours surround generally horizontal alignments of velocity, looking much like layers and lenses of different sediments. Further, the display can be presented in color, with different colors assigned to different velocities. Then the porous zones show up fairly clearly.

DEPOSITION

Seismic sections are now good enough representations of the subsurface that groups of reflections can be used to determine some of the factors in deposition. This is like a geologist's looking at the layering and internal dip on a cliff face and recognizing dune deposits, old talus slopes, alternating sands and shales. Or it is more like a geologist's looking at another geologist's drawing of the layering and dips on a cliff face, with no rock types indicated. This is a part of seismic stratigraphy, the seismic sequence part.

The study of seismic sequences considers a seismic section in segments separated by unconformities. A starting point can be just looking at the section to find parts of it that differ in overall appearance from other parts above and below them. If in one part there are many parallel, fairly strong, reflections, then it is likely that they represent alternating sands and shales in the subsurface. Sands and shales often alternate in thin layers, so there are many reflections, and they usually differ in velocity of sound enough to yield fairly strong reflections. A zone of few reflections, none of them strong, is probably from a massive body of essentially one lithology. Reefs and diapirs are masses, containing few reflections, that extend upward into the seismic times of other reflections. Dunes or other wind deposits may produce many small arcs on a section. Sediments deposited into the edge of a sea are in thin layers in the shallow water, thickening as the sea deepens. Where this deposition crosses a shelf edge, the sediments, in cross section, have a sort of "S" shape. In Illustration 8–3 (from NOAA—it and the other NOAA data were collected by various contractors to the USGS), some seismic sequences can be seen.

SHALLOW GAS

A nonexploration use of seismic data is in the detection of shallow gas that may be a hazard to drilling. Small reservoirs of gas, at shallow depths of less than a few hundred meters, are dangerous to drilling operations. The gas may ignite, destroying rigs and endangering people. When it is encountered offshore, it can also create another problem. The addition of bubbles of gas to the sea water can reduce the density of the water, reducing the buoyancy of floating equipment. A seaworthy ship may suddenly not be buoyant enough to stay afloat in the light mix of gas and water. It is important to know in advance about these shallow accumulations of gas.

The main indications of shallow gas on a seismic section are shallow bright spots. If there are any of them on a section at a point that is to be drilled, they should be considered in planning the drilling program for the well. The need for caution depends in part on experience in the area. The risk is significant in some areas, less so in others.

If your area is one in which there have been shallow gas blows, then you may need to make a special effort to find indications of them. The simplest and quickest is just looking through the sections. A more effective way is to have the sections in the vicinity of the location

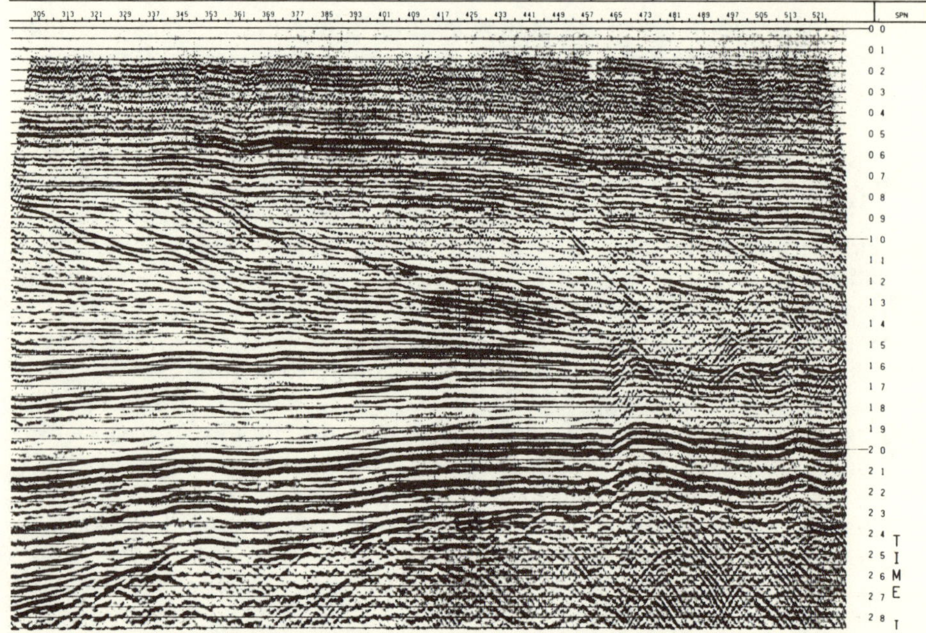

Illustration 8–3 Seismic sequences
(data supplied by National Geophyscial Data Center, NOAA)

reprocessed to enhance the shallow data, and played back as true amplitude sections. A section with bright spots that may indicate shallow gas is in Illustration 8–4.

OVERPRESSURED ZONES

Seismic information can be used in another nonexploration application, the detection and measurement of overpressured zones to direct the use of specific mud weights in drilling a well. Velocity information is the parameter used.

In a porous formation, the rock normally forms a strong framework that supports the overburden, that is, the rock and fluids above it. The fluids in the formation are free to support only themselves. The pressure on the fluids is hydrostatic pressure, equal to the weight of a column of the fluid reaching up to the surface. The fluid may actually reach the surface, at an outcrop of the formation. The fluid that extends through the formation to the surface is usually salt water. The pressure on the rocks is called overburden pressure, and the pressure on the fluids is fluid pressure, or formation pressure.

The rock normally supports the rock overburden, with the fluid supporting the fluid above it, but fluid pressures can in some situations

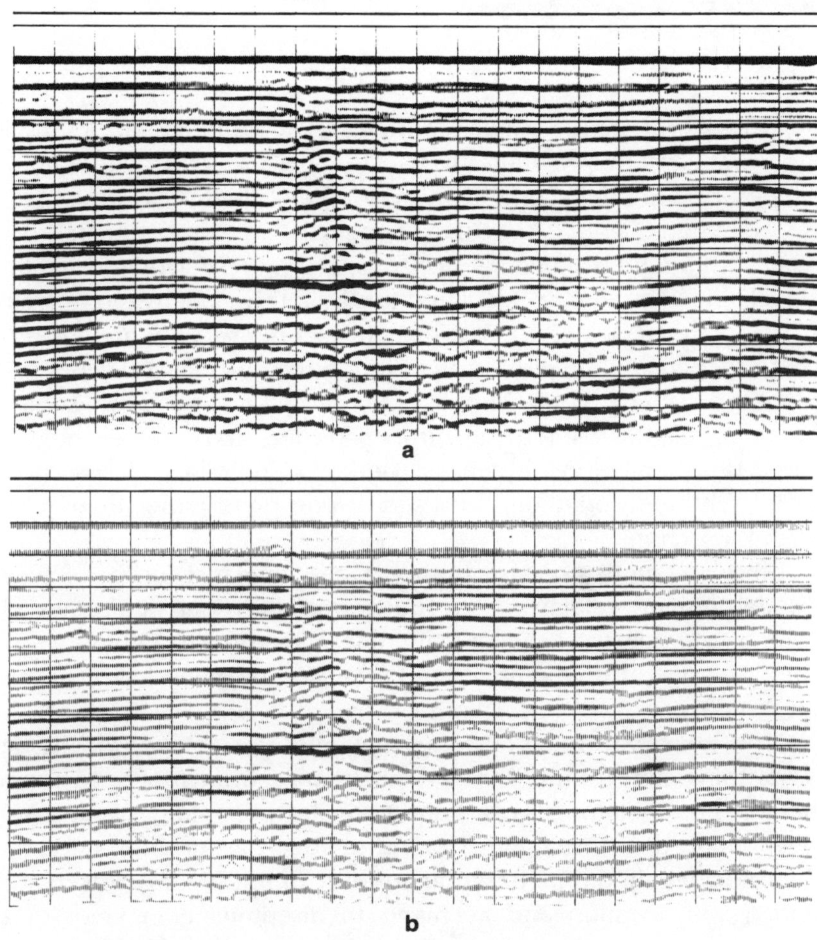

Illustration 8–4 Shallow gas (a) on conventional section and
(b) on true amplitude section
(courtesy Digicon Geophysical Corp.)

be abnormally high or low. Low pressures may occur when the fluids in the formation do not extend all the way to the surface or where there is an outcrop at some lower level, as though the plug had been pulled. High pressures are the more important source of drilling problems. A high-pressure situation occurs typically when, for some reason, the fluid in a formation supports some of the weight of the overburden. This can occur when fluids in a formation are not free to move, that is, are either sealed in or their movement retarded. Then, if some force increases the amount of fluid or if the rock structure is weak, fluid will support overburden, so there will be abnormally high pressure on the fluid. Fluid can even have greater pressure than the total weight of overburden, if some tectonic force is squeezing it against a strong formation.

The overpressured zones produce special requirements for drilling. Mud weight and bit programs are affected. Seismic velocity information, obtained before drilling, can fairly well predict fluid pressures.

Normal moveout can be used to calculate average velocities down to each of the good reflecting horizons. From these average velocities, interval velocities can be calculated for the intervals between reflectors.

The interval velocities, in meters or feet per second, are converted to their reciprocals and plotted against depth as microseconds per meter or foot, on semilog paper. This is a plot in the same form as a sonic log taken in a well. It is a sort of substitute sonic log, made before the well is drilled. With limitations, it can be used as a sonic log. This is another type of synthetic sonic log, but used at one point rather than at each trace of a section.

The limitations concern the crudeness and inaccuracy of seismically determined velocity data. The normal moveout diminishes with depth, so the quality of the velocity determination also diminishes with depth. And of course, multiplies and other seismic phenomena interfere with the data.

Thus, to squeeze the best information out of seismic velocity data, several things are useful. The spread distance from source to farthest detector should be long, say around two kilometers, to get large normal moveout and therefore good data in the deep part of the section.

Velocities should be determined from a number of points near the planned well location and averaged to discriminate in favor of the desired data and against other effects. Velocities should be determined at from 5 to 25 locations.

The velocities should be obtained with greater precision than is customarily used for seismic purposes. Velocities are normally determined in the course of the processing of a seismic survey. But these velocities are designed for economy in covering a large area and are smoothed, to make the velocities consistent over an area. For well prediction, the economics are different, as only one vicinity is of interest and an expensive well is to be drilled.

The interval velocities, plotted on semilog paper as microseconds per meter or foot, will form a straight line of velocity increasing with depth, if compaction alone is affecting velocity. Other factors, particularly lithology and porosity, cause deviations from the straight line.

Overpressured zones, in which the fluid supports some of the overburden weight, or in other words zones with excess fluid, are less dense and more porous, so they have slower than normal velocities.

These often cause more dramatic deviations from the straight line than lithology or other effects. The effect of lithology can frequently be detected as such from knowledge of the area from other wells and from seismic interpretation.

If the plot shows a straight stretch, followed by a strikingly low velocity zone below it, one that is not attributable to the suspected lithology, then this zone may have the same lithology as the straight stretch but may be overpressured. To calculate mud weight to be used for this zone, a straight line is fitted to the straight stretch and then deviation in the direction of lower velocity is measured. This measurement is usually made using a chart that converts directly to mud weight. In general, a separate chart must be made for each geologic province.

We have gone over the main geologic situations you are likely to encounter and some of the special uses for seismic data. A knowledge of these things enables you to interpret seismic data. When you've interpreted an area, what do you do with your results, put them back in the file? We'll explore some more productive uses for those results.

PART **C**

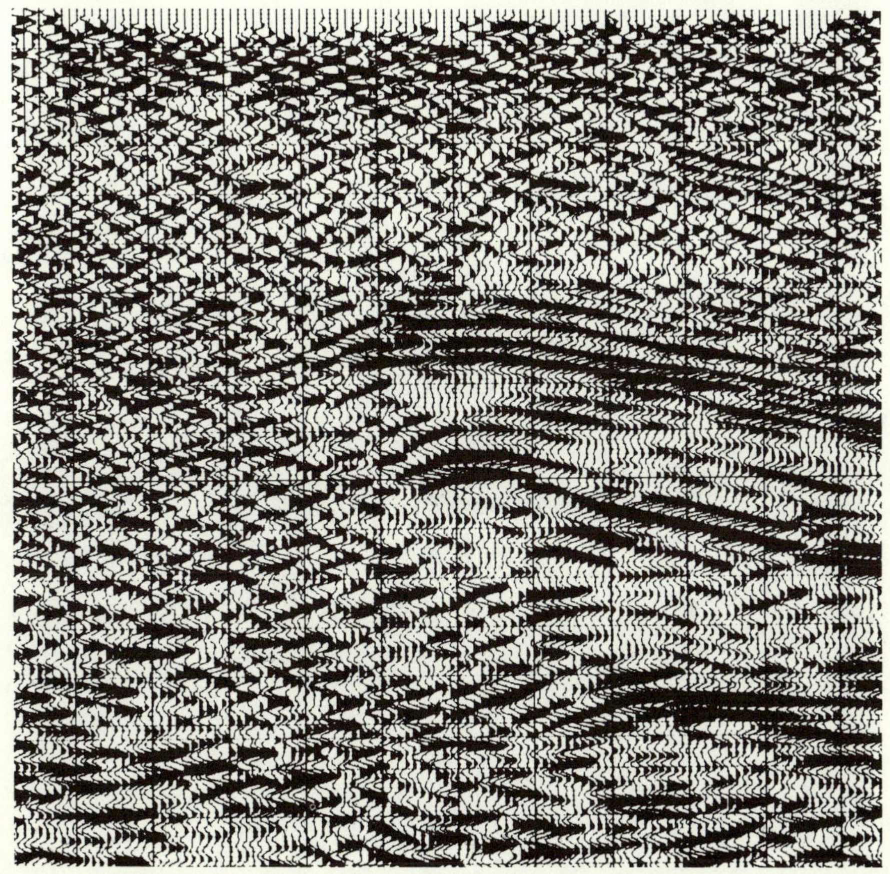

A PROSPECT

Work a Prospect

In interpreting the different types of subsurface features, we want to find prospects and then develop them into producing oil fields. We will take up some special characteristics of prospects and ways of working with them.

LEAD, PROSPECT, DRILLABLE PROSPECT

A seismic prospect is a local area that, for one or more reasons, is thought likely to have hydrocarbons in commercial quantities. There are degrees of likelihood, so people speak of variations of the concept, referring to leads, prospects, and drillable prospects.

A lead gets its name from merely leading you to think there may be a prospect there. It is just a clue, only good enough to justify more investigation, usually in the form of more shooting. A prospect is more developed than a lead so that you feel it has a fairly good chance of containing oil. It may or may not need more shooting, and it may or may not be ready to be drilled. A drillable prospect is one that has been developed to a stage at which the reasonable next step is to drill a well on it. These terms are all rather indefinite, as people use them loosely and as people differ about what stage a particular prospect is in. And other factors outside of the likelihood of finding oil influence the way a feature may be thought of. As an extreme example, if a company will lose its rights to an area if it doesn't drill a well immediately, then a mere lead may be considered drillable.

In general, though, people think of the features in terms of the evidence for oil. This may involve any of the various types of traps discussed in the chapter on interpreting subsurface features. If there is an anticline on one line and there are no other lines nearby, it would normally be considered a lead. With only the one line there is no way to know what happens off to the side of the line. The structure may keep climbing to one side.

But the high on the line does indicate that there may be a local high, either on the line or to one side, so this line is on a flank of it. More investigation is called for. More lines could be shot over and around the feature. Then, if the lines are shot and show that there are lower points most, but not all, of the way around it, the feature can be thought of as a prospect. If still more lines totally surround it with low points, it is a better prospect but perhaps with the highest point not yet established. Later establishing of the highest point by more shooting would still not prove that it contained oil. But that might be a stage at which additional shooting would not make the prospect much better, so the best next step in investigating it could be to drill a well on it. It is then a drillable prospect, one that appears worth the financial risk of drilling.

That example was only an example, and it isn't like all prospects or even all structural prospects. The additional work on some might be, not more shooting, but a gravity or magnetic investigation, a detailed study of well logs by a geologist, or reprocessing of existing seismic sections in the area.

THE "PROSPECTIVE" OUTLOOK

A large part of your job is to find prospects. How do you go about it? As you pick and as you contour, look for places, sometimes not your best decisions, but alternate choices that might be prospective. If you see a place where the reflection might go higher or a hint of a reef or fault trap, stop and think about it. Then you may want to either take the more optimistic choice or at least make a note of that possibility for future consideration. This is not the time to decide which choice is more likely to be correct and ignore the other. If the optimistic alternative looks poorer but is really correct and there is an oil field there, you would cut yourself off from discovering it. If you take the poor but optimistic route and it is incorrect, you still have plenty of chances to find out that it is wrong.

This is the essence of the often expressed philosophy that explorationists should be optimistic. An interpreter who makes a too-optimistic wrong choice may waste some money in interpreting time, additional shooting, even drilling. But an interpreter who makes a wrong choice that is too pessimistic loses the company a chance to find an oil field. A good oil field will bring in many times the money that would be lost in some wasted exploration effort. And an optimistic choice in seismic interpretation is not necessarily a commitment to a great deal of expenditure. It is bringing a possibility up for consideration.

DON'T GET TOO MUCH FROM THE DATA

Seismic sections were obtained at much expense by people using the highest technology and sometimes expending great effort. These sections are the one thing we have to show for all that, and they are the main thing we have to work from in doing our interpreting. We need to get all the information we can from it, right? Well, yes, in a way, and no, in another way.

Much of interpretation is an effort to get all the information that can possibly be extracted from the sections. But there is also a danger of getting supposed information from the section that is not really in it.

If you try too hard to see lineups of reflections on a poor-quality section, you are likely to start imagining reflections and then, in a next step, believing them. That is the point at which you are getting too much "information" from the section. Then it is time to stop, back up, look things over with a detached mind. If that makes you fairly sure you were just imagining the reflections, then erase your picks. It may then be best to come to a different kind of conclusion, a perfectly valid one—that you can't interpret that reflection.

One time, with someone else, I was trying hard to make some picks, of any reflections at all, on a very poor section. We had the section down on the desk between us and were peering at it from opposite ends. Then we started to see a dim reflection, hovering around a time of one second. Squinting and advising each other, we managed to get some distance on the section colored. We called it the "one-second reflection." Then we started to notice that the thing we were marking was very flat and right at one second. Finally, we realized that the heavy timing line at one second was enhancing the noise enough to make it look like a weak reflection. We had been picking the timing line!

Contouring also can yield too much "information." Contouring is always a guess when you get away from a seismic line, and the farther away you are from control, the more you are guessing. Be somewhat careful about contouring in the blank spaces far from the data. Reasonably close in, you should guess. In your area you can do it better than other people, so don't be ultracautious, not daring to contour any distance from control. But out in the large open spaces, admit that you don't know.

SECRECY

Seismic data is the property of the company that has it shot, to use or sell or trade or whatever. Usually the data is very secret. The reasons for the secrecy have to do with competition for oil.

In the United States the mineral rights belong to the owner of the land. Except for public lands, the land is mostly owned by individuals, often farmers or ranchers. For an oil company to drill a well, it must first lease the land from the owner. This makes leases small in area and individually purchased. The leases expire at various times. A company's leases tend to be scattered in with other companies' leases and with unleased land. In these circumstances, seismic data must necessarily be kept very secret. After a prospect is interpreted and recommended for drilling, specialists in buying leases buy what leases they can on the prospect to add to those the company may already have.

It is good economics, when you discover oil, to have the whole field leased if possible. Even then, there are plenty of chances for the field to cost more than its production is worth. Obviously, another company in possession of the same data has a good chance to get some of the leases without having had the expense of the exploration. Prospect information must be guarded like the oil fields it represents. Oil fields are like money—banks don't leave it lying around on counters.

In an area where oil development is by concession or production-sharing contract, the contracts are for a limited period, usually with the provision that parts of the area must be relinquished in stages through the term of the contract. In this case, other companies' knowledge of your area would give them an advantage in bidding against your company when the areas are offered again.

Another reason for secrecy is to allow a company to have information to offer to trade for information from other companies. If copies of the data had already been passed around, there'd be nothing to offer for the other data.

A seismic contractor may shoot some regional lines, or some lines in an area that is to be made available for bidding, and sell copies of the data to any companies that are interested in the area. The need for secrecy by the contract company is to keep the data marketable. If copies were freely available, no one would pay money for them. Parts can be shown to prospective customers or even published, with exact locations omitted. Companies buying the data are required by contract to keep it confidential.

One situation that calls for less secrecy is in lines shot by government or educational institutions for regional geological research for the purpose of making the data available to any interested parties. Even this information, though, may be considered confidential until it has been assessed and is ready for release.

As I said before, exploration information is the property of the organization that had it shot. An interpreter or anyone else handling the data is entrusted with it for safekeeping. The data is to be kept totally secret unless and until management makes it clear that some specific, more open handling of it is permissible.

A sidelight on this occurred when I was asked to answer a questionnaire about a seismic technical assistant who had applied for a job on a police force. One question asked if I would trust him with some specific large amount of money. I was able to answer that we already had been trusting him with things of more value than that amount. I was referring, of course, to the seismic data he had been working with.

WORK EFFECTIVELY ON A PROSPECT

One of the things that makes a person an effective interpreter is a tendency and an ability to look at different aspects of a situation. When you are given an area to interpret, your main effort will go into trying to pick the sections correctly, so the loops are tied, features recognized, and maps contoured correctly. You might feel that you will do the work best by paying close attention to it and not being distracted. But there are always other things to be considered in the area.

Your interpretation can be more effective if you know something of the geological background, of the company's reasons for having an interest in the area, etc. This does not mean that your picking should be made to fit some preconceived idea or tailored to finding a prospect in a certain place. But, for instance, if you find a major fault, it's good to already be aware of the trend of faulting in the area. Then you can decide whether your fault fits the trend or is an anomalous fault that goes against the trend of most faults or may even be a clue that the "known" trend may not be correct.

If you find what looks to you like a prospect, you should not only bring it to people's attention but, if it is ignored, persist in bringing it up when proper occasions arise. However, at any stage in its development, you may find evidence that makes it less prospective than it had been. Be open-minded about it and also don't keep the information to yourself. Once you have good evidence that it is not a prospect, you should point that fact out to your boss and be ready to drop the feature from your activities. Then you can spend your time on something more worthwhile.

When you are investigating a possible prospect, try to think of all the factors that might be important to it. Get out all the sections that cross it and maybe some of those around it. Pick them, this time without consideration for how they fit the rest of the area but just concentrating on the lead. If your lead seems to be the same kind of feature as some already known field, get out any sections over that field that are available and compare them with your lead. Make a new map, of just the immediate area of the lead, still ignoring how it fits the rest of the area. If you do have a prospect, you may later have to do some reworking to make it fit, changing either it or the other parts of the area.

In some situations a prospect might not be found at all until it is contoured optimistically, to see if it can possibly be contoured as a prospect. This is an important part of exploration—to try to find everything that might be a prospect, then investigate further, wiping out some of the possible prospects but confirming others. After a prospect is found, and appears fairly good, it is sometimes worthwhile to contour pessimistically, to see if you can destroy the prospect. If it can't be destroyed by any reasonable contouring, the prospect is supported. But, if it can be destroyed, this does not necessarily mean that it is a poor prospect. Accept evidence that wipes out a prospect but lean toward the positive. If the negative evidence doesn't conclusively destroy it, keep investigating the feature or, at least, keep it in mind. Try these two ways of contouring in Exercise 9–1.

When you do work up a prospect, you will probably need more shooting to confirm it and develop it further toward drillability. This is likely, as areas are usually shot with the lines too far apart to thoroughly define prospects. A new program will need to be planned and shot. Detailing a prospect will be taken up in the chapter on program planning.

It is possible that a lead may quickly develop into a prospect. There may be enough evidence in the form of sections, wells, etc., to make it a prospect, even a drillable one. If so, fine, you're lucky. More often, though, there is a lot of time involved. The lead may call for more shooting or for trading to get seismic sections or well data from another company. It may not even be ready for that, but just for keeping in mind so you can be aware of the influence on it of things you learn later. In these situations, years may go by before a feature is drilled. That's fine, the company may have a number of these slowly developing prospects so that, every so often, one gets to the drillable stage, and some of those are discoveries.

Exploration for oil, like science, is a struggle to learn things about the world that are difficult to learn. It is also competitive. If your company doesn't find the oil in a place, another company will. In the U.S., companies may have leases in the same area. In countries with concessions or production-sharing contracts, there are usually requirements that parts of the area be relinquished at intervals. Later, others may find the fields you missed.

You need to try to be better at finding oil than other people. You can be as good, or somewhat better, by being very careful, using all the knowledge you have. To be a lot better, you not only should have an optimistic outlook, but you should also use imagination.

Once oil has been found in an area, then more fields of the same type can be found, by looking for the same kind of feature in the same geological conditions. This finds a great deal of oil. But a person with a new idea, maybe seeing and investigating a different phenomenon on a section, may find the first field of a different type. That person and that person's company then have the best chance to find more fields of that type.

If you do the regular things, looking for the type of prospect that is conventional for an area, you can be sure that that is what other people are looking for, too. That's good, you'll find oil. The known tools should be used. But, also occasionally try something different. Maybe look for prospects in places that do not look prospective by the conventional ideas. You could come up with something.

ALL THE HELP YOU CAN GET

In looking for oil, we are trying to learn things about nature. We should get information from any source there is, except, of course, for stealing from other companies.

This acquiring information is quite different from the situation in a school, in which you are being tested on your individual knowledge. In a test in school, or in homework, the purpose of the endeavor is to determine how much you know. This isn't determined if you present someone else's knowledge as your own. In this real world endeavor, though, you can—and it is your duty to—obtain and use all the information you can find that applies.

When decisions are being made about prospects or about wells, it is useful for people to think about all aspects of the situation—seismic, geologic, economic, engineering, access, rig availability. This bringing together of information from different specialists is primarily the duty of management, but an astute person in any specialty may have a fact to contribute or may think up a question no one had considered. It is surprising what good points may be brought up by people with different types of expertise and different ways of looking at things. You can profit immensely from them.

Read articles about the area and similar areas. If you see an interpretive idea or a geological fact in print or if someone tells you of a technique that worked in another area, use it. Don't tell yourself that you should do your interpreting somehow on your own, refusing to use some data because you didn't get it from the sections. You are pretty much on your own anyway, even with all the help you can get. You are trying to find some oil that nobody has found before. That is something like research, and like new discoveries in research, all the knowledge and techniques the human race possesses haven't found that oil yet. Afterward, people won't be nearly as interested in just how much of the thinking was your own as they will be in whether or not you found oil.

BEING CAREFUL AND TAKING CHANCES

Human beings, by nature, are careless. A large part of the raising of children is teaching them to be careful about certain things. Similarly, people starting to work in geophysics have to discipline themselves to be careful about data. In school they are taught mathematical and physical principles and are given a little practice in working out problems. However, the only thing that happens when the answer to a problem is incorrect is that a grade is reduced. In hunting for oil, much more depends on the work being correct.

An example is plotting data on a map. If one digit of one number happens to be incorrect, the datum at a shot point can appear to be milliseconds or tens of milliseconds higher than it is. Contoured, this can look like a prospect. If people followed it up uncritically, a well might be drilled on it. Or if things didn't go that far, just some detail shooting might be done there unnecessarily. Or management might be stirred up and then let down. Or you might be embarrassed. The first of these is a fantastic waste but is unlikely because other people would be skeptical. Each succeeding one would be less harmful, but even the last is undesirable and might affect your career.

The way to avoid this is by checking. When you work with data, you must check yourself frequently. You must learn to recognize abnormal data and check them even more. Also, your superiors have trained themselves to be suspicious of unusual-looking data, so the first question one of them may ask is likely to be one you hadn't thought to check. When you're asked one of those questions, believe me, it makes you feel foolish and kind of incompetent. I've been there. All experienced geophysicists have. In fact, they never quite get away from it. A skilled and experienced interpreter can check everything thoroughly, only to have the next person who looks at the prospect ask a question that destroys it. The trick is to keep these times to a minimum and, when they do occur, to accept them gracefully and be glad the problems were intercepted when they were instead of going further and costing more time and money.

You should freely admit it when you are wrong. But, of course, you can't afford to be in a position to need to do that too often. So you have to train yourself to careful work, with thorough checking, particularly of things that look like prospects. Most people in geophysics learn this lesson and become properly careful. Good. But then another problem crops up.

In looking for oil, you need to be willing to take chances. All that checking was to try to have your data and contours as correct as possible. But the conclusions you draw from the finished interpretations are less exact. They tend to be just impressions or even guesses. For example, the interpreter may finish mapping and then think, "Part of the area looks high, but the data is poor right there, and the fault may bring it down instead of up." Or, "I believe there is a reef there. It looks like a drawing of a reef, but not quite like any reefs I've seen on seismic sections." What do you recommend: drilling, more shooting, dropping it, or what?

As interpreters become more experienced and their careful work and judgment are trusted more, they get more involved in the chancy decisions. Some people, when they reach this stage, retain the careful attitude and carry it over from data to drilling decisions, refusing to take chances. How do you train yourself to be careful with data but willing to take chances when necessary? Normally, a careful person doesn't like to take chances, and a chance-taker doesn't have the patience to be careful. A way around this is to be careful with data and, in looking at a prospect, to try to think up all the factors that may affect it and then to weigh them, considering not only the risks of drilling but also the risks of not drilling.

People who regularly take negative views of prospects are motivated by caution—they don't want to cause an expensive dry hole to be drilled. But there is another kind of caution to be used, too—being careful not to cause a good oil field to be missed. When you think, "How will I feel, and appear to others, if the company spends all this money and drills a dry hole?" think also, "How will I feel, and appear to others, if I cause us to not drill, and someone comes along later and finds oil here?" That's a worry many of the cautious people haven't thought about, a factor they aren't cautious about.

CHAPTER 10

Show It to the Boss

After you have interpreted an area or prospect, you are only partly through with it. If you and everyone else leave it at that, it has just been an exercise in interpretation, useful for developing your skills but of no other benefit to the company. It must be presented to other people who can decide what action is to be taken on it.

YOUR BOSS

If you are working in a geophysical department, your boss will be a geophysicist, too, and will probably be more experienced than you are. To show the boss what you've found, no special presentation is necessary. There will be questions, so it's helpful if you can have the answers ready beforehand. It's a waste of the boss's time to have to wait while you find out if the line was shot 1,200% or 2,400%, or if there's a better version of the critical line, or if a well has already been drilled on your prospect.

You don't need to be slavish about having everything ready before you show your work, though. Sometimes, especially if the boss happens to walk in just when you've discovered something new and exciting, it's good to say something like, "Hey, look at this. I haven't had a chance to work on it yet, but it looks interesting." Bosses like the excitement of discovery, too, and may add some good thoughts to yours.

In looking over your work, another geophysicist does best with sections, preferably not colored too heavily, and maps that have the data legible enough to judge the contouring. Work sections and work maps are often best for this purpose.

An entirely different situation prevails when a prospect is being shown to management.

YOUR BOSS'S BOSSES

After an interpretation is made, the results will probably be shown to other people by your boss, maybe with your help, or maybe just by you. These other people, probably managers, with specialists in other fields available, will compare it with other types of information, weigh it against other factors. They will probably make some decision as to what to do about it—whether to drop the area as of no interest, do more shooting, gather other types of information, or drill a well on it. In this process it may be presented to others several times, perhaps in several degrees of polishing.

Managers mostly are people who are not trained in geophysics. And anyway they usually don't have time to go into detail on your pet prospect. They want to know things like what kind of prospect it is, how enthusiastic you are about it, how carefully you have worked it up, how competent you are, what other specialists think of it. From this general kind of information they can form opinions to act on.

To show management a prospect, it is usually best to:

☐ Make clear, simple displays,

☐ On sections, color important picks boldly,

☐ On maps, use bold contours—legibility of data isn't so important,

☐ Know the background information—about leases, wells in the vicinity, etc.

The simplifying is because they won't want details but will want to understand the overall situation and also because, in a meeting, people may need to see the important features of displays on the wall from across the room, where they can't see the detail at all.

You were hired to convert sections to an interpretation, something understandable to people outside your field about the likelihood of finding oil. Part of your job is to simplify what you have understood and to present it so other people can understand enough of it for their purposes. There is a kind of snobbishness among technical people that causes them to expect others to understand their technical language. But it isn't good to treat your bosses that way. Instead of thinking you know a lot of technical language, they may just decide that you can't explain things very well. They may even have picked up a fair amount of geophysical jargon, but that doesn't mean they understand it all.

How do you present technical information then? You have made decisions, some of them for fairly subtle reasons. You will have to leave the subtlety, or most of it, out of your presentation. It is all right to let people know you can't be sure of something, if that is the case, but they won't be able to follow you if you go into detail about why you aren't sure.

MEETINGS

For the meeting, if there are just three or four people, you could just put your maps and sections on the table for them to see. Or they might see them better if you put them on the wall. And certainly for a larger group, it is necessary to display things on the wall for all of them to see.

For display, a section should have only a few horizons marked on it, and those should be colored strongly enough to be seen easily from across the room. For your interpretation to show up well on a wall in a meeting or just for you to look at and draw interpretive conclusions from it the coloring must be bright. Almost any time a person is shown a brightly colored interpretation, some comment is made about the influence the bright colors have on the viewer. They lead the eye so that it is very difficult to disagree with them. This means, then, that you should use the strong colors to make your interpretation visible, but that it is also sensible to have an uncolored copy of the section available for the people who wish to make up their own minds about the interpretation.

Sometimes, for meetings, it is useful to have some extra-large-scale sections, maybe twice the normal scale, made to show details clearly. This is particularly true in areas of small or subtle features.

A map needs to have the contours, or the main contours, strong enough to be seen clearly by everyone at the meeting. It may also be useful to color the map. There isn't any one way to do this. Just highlight whatever you think is most important for people to see on it. You could fill in the closed high areas with one color and maybe stress some main low trends in another color. Reefs, salt domes, company leases might be good things to make stand out in color. There is a convention of coloring, not necessary to be followed, but sometimes useful because people seeing it are familiar with the convention. In it, red represents gas, green is oil, blue is water. On a seismic map you usually don't know whether there will be oil or gas but can use the convention by making the highs red or green and the lows blue. Be a little cautious though. If you know that a person you are showing maps to is colorblind for red and green, then something else like yellow highs and blue lows would work better.

There is a convenient way to color areas on a map smoothly. Use colored pencils that will smudge easily. Then go over an area you have colored by rubbing it with a facial tissue wrapped around a pencil. This will smooth out your colored pencil strokes.

In Exercise 10–1, color a map in such a way as to make the important things stand out.

The data on the map don't need to be so visible. Some companies even leave data entirely off their final maps, showing only shot points and contours. There is a trap in this, though. If you are asked some question in a meeting, you might not be able to get the answer from the map. Also, anything else you think you might want to know in a meeting you can put in small print on the map or section. It doesn't clutter the view of others but is available to you to use as reminders if you need it or even for them if they want to go up close and get involved in detail.

When you are asked to present something at a meeting, it is important to take what you will need into the room. First, select the main items, maybe one or two key sections and one map. Then also get whatever other material you might need if people quiz you and insist on seeing more evidence than you have shown them. This extra material makes a stack beside you in the meeting room but may keep you from having to hold up the meeting while you go back to your room for something. Try to have the main items in good form for display on the wall.

If you are still at the work map stage it might be good to heavy up the key contours with a soft pencil that can be erased later as your interpretation changes. If the map has only pale contours, it may look like a blank sheet of paper from across the room. These meetings can be important. It might be decided to give your interpretation project priority—or to abandon it. Do what you can, quickly, to get things readable—not pretty, the people understand that work maps aren't supposed to be pretty—but so they can see what you are showing them. If there isn't time, of course, use what you have.

MONTAGE

An excellent technique for displaying a prospect to management or to a potential drilling partner is a montage. The word is used in art to describe a pasteup of different things onto one sheet. The idea is the same in oil exploration. It refers to putting the main information about a prospect on one sheet of paper. It supplements or replaces a bound prospect report that some companies use.

A conventional prospect report is in the form of letter-size pages put together in a binder, like a thin book, with maps, sections, and the like folded and placed in a pocket inside the back cover. When someone wants to show the prospect, the binder is brought out of the file, the enclosures are unfolded and stuck on the wall or laid on a table, and everybody tries to see everything. After the meeting it's lucky if all the parts get back together. And of course if they don't, then the next meeting on the prospect is a real circus of hunting up things.

One time a weekly progress report from a seismic crew, just a map and one or two typed pages, had been sent to a head office. The head office telephoned to ask what was meant by something on the map. "But that was explained in the report." "Oh. But we file the reports on a different floor." "Why do you do that?" "It's just the way we're set up." A montage is some protection against that sort of thing.

To display a prospect in montage form, get a big sheet of tracing paper or film and tape onto it transparencies of the items necessary to give the idea of the prospect and maybe to persuade people to drill it. These items are the highlights of the prospect, not a complete report on it, and involve all aspects of the prospect, not just the geophysics. Making a montage is a combined effort of geologists, geophysicists, maybe production people.

There isn't a fixed list of the items because what is needed depends on the prospect. The items typically include a well log cross section, a seismic map, a couple of key seismic sections, an index map, a typed geological discussion, a well prognosis. There may be more or less of some items, some may not apply, others not mentioned may be included. A picture of an outcrop or anything else pertinent might be included for a particular prospect. In every case it is necessary to include a title block to say what the montage is.

Once everything is arranged and taped on the backing, a diazo print can be made. After the prints are made, some reflections or other features on them may be highlighted with color. Then getting organized for a meeting is simple. Just put a print of the montage on the wall. That's it, except for the usual backup stack of sections, etc., beside you. If the attendees want something to take back to their offices, hand out folded copies of the montage.

It's refreshing to see many questions answered by referring to different spots on this one piece of paper. Nobody forgets to bring an important item to the meeting or loses it out of a report. And a request for additional copies is satisfied by making more prints, if uncolored copies are adequate for that purpose.

There are a few tricks in preparing the master transparency. I glibly said to tape the individual transparencies onto it. In the printing process they will all be given the same exposure, as the whole montage is printed all at once. If some parts are dark and some are light, it won't be possible to make a good print of the whole thing. The way to best resolve that problem is for each item to have as much contrast as possible—with very opaque lines on a very clear background. Photographic film is superb—dense lines of metallic silver on highly transparent film with either clear or matte surface. However, this photographic copying is expensive. Transparencies of maps and sections are normally made on the cheaper and more convenient sepia materials, which do not have as much contrast as photographic film.

A good montage can be made if some precautions are taken. The transparencies should be as few steps from the original as can be arranged, to minimize the loss of detail with each successive stage of copying, and should have just enough background color to ensure that no detail is burned out. Sepias should be "reversed." Then the brown lines are on the underside where they will be in direct contact with the printing paper. And for the lines to touch the printing paper the sepia must be taped on the underside of the large sheet the montage is built on, that is, taped facing it.

The typewritten parts are most easily made transparent by copying on transparent film on an office copier. The film is sold for making slides for overhead projection and is quite inexpensive. Ordinary drafting film or tracing paper can be cut to size and run in many copiers.

Some people, after putting the pasteup together, like to make a sepia of the whole thing to use as the printing master, so there won't be pieces coming unstuck during printing. This is handy, but it's another generation from the original, so there is a little loss of sharpness. Also, the remarks above on putting things on the underside don't apply if such a master sepia is made. The arrangement must be designed for this use, with face-up sepias and other transparencies put on top of the big sheet. Then the sepia made from the paste-up is a reversed sepia.

All this is a lot of taking care, but it's well worth it when you can just whip out a clear, legible print that includes all you need to display on a prospect.

WRITTEN REPORT

After a seismic interpretation is finished, it can be useful to make up a written report on it. The report serves two purposes. It is used to tell people the results of your interpretation. For this purpose you need simplicity, at least in a summary, that will tell managers what you have learned about the area. The other reason is to preserve the information gained in the interpretation. For this, details are needed. You are writing the report to yourself as well as to others. Your memory will not retain the problems, the stage your thinking has reached, what you think should be done. A year or so later when someone asks about the area, you need to be able to get out the report and refresh your memory. Or if someone else is to work on the map, maybe after you have been transferred to another office, that person should be able, by use of the report, to not have to duplicate your work, but continue from somewhere near where you left off. Some companies require these reports when interpretations are finished.

For all these needs a report on an interpretation should have about everything that might sometime be needed:

☐ General information, like dates of shooting and interpretation, contractor, field crew, data processor;

☐ Where the data for the area is stored;

☐ Some examples of the data;

☐ Methods of interpretation used;

☐ Problems and, if possible, solutions;

☐ Maps of horizons, intervals, etc.;

☐   Conclusions drawn from the interpretation;

☐   Recommendations about action to be taken.

The conclusions and recommendations in particular should be given, or summarized, in a form understandable to people who are not geophysicists. Details about the data and interpretation, though, should be given as completely as may be necessary for future reference, so they can be in technical language for the use of other geophysicists.

The traditional way of writing a report is to lead up in the course of the report to the conclusions and then to the recommendations, so the conclusions and recommendations are some of the last items in the report. Another way is to put them at the first so those who need only that information can find it quickly and so people can read those parts to decide whether to read more of the report. Either way is satisfactory, as long as the conclusions and recommendations are clearly labeled and easy to find.

The maps and sample sections are usually folded and put in a pocket on the inside of the back cover of the report. Then, to look over the report, a person can get them out, unfold them, and look at them while reading the text of the report. The sections and maps are usually large. They take up a lot of room and call for some sorting to find a certain one. They might be misplaced, put on the wall for further study, borrowed—and not put back in the report. Then the next time the report is needed, it is not complete.

The solution to this problem is to make reduced copies of the large items and bind them in the report as pages. They can be the size of a page, or the height of a page but horizontally longer and gatefolded or accordion folded. There can also be the larger copies in the envelope, or not, depending on the situation. If the reduction for binding makes the lettering too small to be read, then there should also be larger copies. If the large ones get lost, there will at least still be the bound reductions. People usually prefer to look at the reduced maps and sections. It's easier and quicker to turn from one to another than to handle the large ones.

ARE YOU SURE IT WILL PRODUCE?

When you recommend a prospect for drilling, what would you say if someone asked you if you were sure that the well would produce? There is a simple answer: "Of course not!" We can never be certain that a well will produce.

Many people, some of them, unfortunately, working in exploration, seem to feel that a prospect must be almost guaranteed to produce in order to be worth drilling. That's a pity. We don't have guarantees in oil exploration or even in development of discovered oil fields. A development well surrounded by producers can be dry.

The figures quoted for many years for success in drilling exploratory wells for oil were usually somewhere around one in ten. That is, on average, nine dry, or less than commercial, wells were drilled for every commercial field discovery. The industry's being wrong 90 percent of the time was worthwhile only because a discovery was so productive that it more than paid for all ten wells. The odds have improved recently but are still against us. Someone in an oil company must be willing to take chances on drilling wells knowing most of them will be noncommercial. That person must try very hard to get the odds as favorable as possible. If the exploration is done well the chances for that company might improve.

Some people in exploration become what a friend of mine calls "ninety percenters," referring to the old estimate. With the odds, all they have to do to be right most of the time is to consistently oppose drilling prospects. These "explorationists" say, "No, it's not good enough to drill at this stage" or "We need some more lines over it" or "Let's think about it awhile." If the well is drilled against their recommendation and is noncommercial, they are in a position to remind management that they had recommended not drilling it. They can be right when others were wrong most of the time, which, in most activities, is a pretty good record.

However, there is a counter to this usually being right. People in exploration may say that someone "hasn't found a drop of oil." This is a worse condemnation than saying that the person drilled some dry holes.

DEVIL'S ADVOCATE

The old expression "devil's advocate" refers to a person who deliberately takes a side in an argument, not from believing it, but in order to bring out the best arguments from those on the other side and in general to get at the truth. Anyone who is in charge of other exploration people, or even just dealing with them, finds this position useful in bringing out the positive and negative features of a prospect someone is presenting.

When you work up a prospect, it will help you to assume that position yourself, seeing how you can answer difficult questions. Also, you can enlist a co-worker to think up questions and negative factors for you to consider. Both of these activities will make you more ready to defend the prospect later or, perhaps, may show you that it is not a prospect.

There is a danger, though, that people will take this devil's advocate game too seriously, forgetting its real purpose and playing it to win. Any person in exploration, particularly one in authority, can seriously hamper exploration by taking the devil's advocate position, not to bring out good arguments, but to kill prospects.

SCIENCE AND EXPLORATION

In scientific investigations skepticism is essential. A scientist does not proclaim a law of nature until the idea has survived the tests that scientist and others may apply. The evidence must be so good as to almost constitute proof. This skepticism is the basis of science, distinguishing it from rationalization, tradition, philosophy, and other forms of guesswork. That skepticism is reserved for proclaiming laws of nature. It is not used in deciding whether or not to try an experiment. Science develops from trying any experiments and then being skeptical about the results.

Oil exploration is similar. The part of oil exploration that compares to the experiments is the work itself—the field geology, seismic work, drilling of wells. The question of whether to drill a prospect is not analogous to asserting a law of nature but is more like conducting an expensive experiment. A scientist does not feel that it is necessary to prove that an experiment will give the needed answer, but decides on the basis of feeling about the likelihood that it will produce useful results, the money available for it, etc. Similarly, in oil exploration you do not need proof before drilling a well. You can treat the drilling as one of the exploration tools, an expensive one but still an exploration tool. If you waited for proof that wells would produce, no wells would be drilled. At that rate, we know how much oil would be found.

Drill the Prospect

When you have interpreted a prospect and it appears to be worth drilling, you can then pick out a place to drill to test the feature. You choice may not turn out to be the final selection, but your choosing a location helps to get things started.

PICK A LOCATION

There are so many important factors in the selection of a location to test a prospect that ten specialists might come up with six or seven different locations. We'll consider some of the likely locations (Illustration 11–1).

Highest point. The highest point of an anticline is the spot most likely to contain at least some hydrocarbons. Even if the field is not large enough to be commercial, it is useful for further exploration in the area to know what is trapped

INTERPRETING SEISMIC DATA

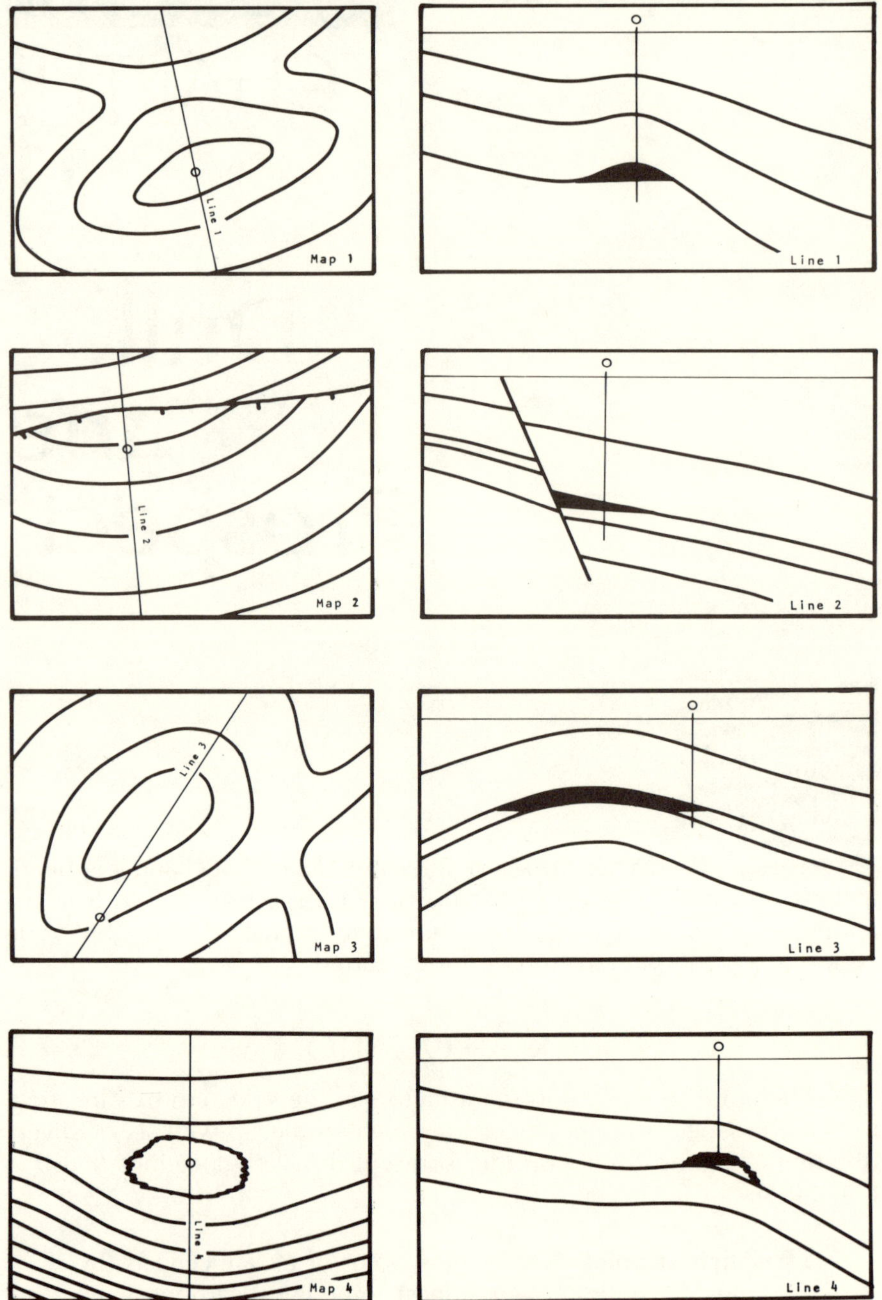

Illustration 11–1 Drilling location

there. And if there is sufficient oil or gas, then production can drain the very top, not leaving behind any attic oil, updip from the highest well.

Away from the edge. If the high point is central to the feature, then it is away from the edge; but a fault trap might be highest right at the fault. Aiming for that point is risky. If either shot point or well is mislocated or if the well deviates or if the fault on the seismic section could be interpreted at a different point, then there is a chance of reaching the pay zone on the wrong side of the fault. Similarly, any areally small feature could be missed if not drilled near its center. And if the location happens to be offshore, there is the added inexactness of finding the location of a shot point that is not marked on the ground.

Flank location. A position down on the flank has a chance of finding an oil-water contact and thus giving an idea of the size of the field. Also, if there is gas at the crest of the feature and oil lower down, then the flank location is more likely to find the oil. However, it also has a chance of missing the accumulation altogether. The usual rebuttal to that is that, if the field is too small to produce on the flank, it is too small to be commercial, and that it is good to learn that on the first well. But, the argument assumes that the mapping is exactly correct—or totally wrong. It almost certainly isn't either of those. I feel that the flank position should be saved for a second well. Prospects are elusive enough and difficult enough to interpret without throwing away one that might exist and be commercial but that is not shaped just as mapped.

Specific geological location. There are many special conditions in the various types of prospect. In drilling a reef, a central position might not be good if that position happened to be in the lagoon. The point of greatest reef thickness might then be more prospective than the highest point. In a stratigraphic trap, thick and porous parts of the formation could be the most prospective spots.

On a seismic line. This point applies to all the foregoing location situations. The area has been shot and a prospect found with the seismic data. The map was contoured on the data. The contours look nice and complete on the map, but they are correct only where they cross the seismic lines. Everywhere else, they are guesses. They may be pretty good guesses in some places, especially near the lines, but they are still guesses. A well located on the basis of that data should be located where there is data, not off to the side somewhere. Then you can make predictions about the well more reliably, predictions like the depth to a formation and that there should be oil in it. A second reason for locating the well on a line is for later use, after it is drilled. If the well is not on the line and if it doesn't fit your expectations, you may always wonder if there was a fault or lithologic change between well and line. You need for the well to be on the line so you can get full advantage of the new information provided by the well.

Use these principles to select some well locations in Exercise 11–1.

Try to consider all the factors in the particular situation you have in selecting a location to be drilled. Then, when other people get in the act, you will have some ideas to compare with theirs. Your location may not survive the meetings, but if your prospect does and if the location chosen does not appear to be one that would fail to find the accumulation, fine. If the others argue about where to drill your prospect, rather than whether to drill it, it may be a good time for you to sit back and enjoy the discussion. The situation is like some advice about selling. When the prospective customer picks up the product and starts playing with it, that's the time for the salesman to stop talking. Bosses like to have some part in a prospect. Let them. You've had the fun of working up a prospect and seeing it proposed for drilling. And neither one of you knows yet whether it will be cause for celebration or a dud.

SURVEY FROM A SHOT POINT

There is a subtlety in surveying well locations that is important to the discovery of oil:

A well based on seismic data should be located with reference to shot points, not to geographical coordinates.

Here is what often happens in companies. Surveying is a part of the process of obtaining seismic data. The results of the surveying are plotted to form a shot point base map. The interpreting is done on copies of this map. A well location is selected on the map. The coordinates of the location are read from the map and put on the well recommendation. A surveyor is sent out to mark that position on the ground, with a stake on land, a buoy at sea. A mislocated well might fail to find an oil field, so a strong effort is made to be sure the surveying is done correctly, that the exact position in the well recommendation is staked.

Because of the expense of drilling a well and the risk of a wrong location, a careful and, if need be, expensive operation is warranted. This degree of care takes considerably more time and expense than can be justified for each shot point of a seismic survey, conducted just in the hope of finding a good place for a well.

Wells are located more carefully than shot points. But the well locations were chosen from shot points. What if the seismic line is mislocated somewhat? What if the coordinates were read incorrectly, maybe read from a stretched print of the seismic map? Do we want the well in a "more correct" position than the shot point we hope to be drilling? The answer to that one is obvious. We want to drill that shot point, not some arithmetically correct match to a set of coordinates read off a shot point map.

How are procedures to be arranged to do this? If the location is on land, there may be ways to find traces of the shooting on the ground. A tag on a bush with a shot point number, a pile of cuttings beside where a shot hole was drilled, the "footprint" of a vibrator, a bench mark put down by the crew surveyor may permit a well location surveyor to find where the shot point is. A tag or bench mark may indicate another shot point on the line, so the well location surveyor can count piles of cuttings or just measure distances along the line to find the desired shot point or a good estimate of its position.

A necessary first step in getting the well drilled on a shot point is on the well recommendation form itself. The shot point should be given as the location to be drilled, with the coordinates only listed as secondary information, for example, "Shot Point 357, near Lat. __, Long. __." If this is not made clear on the form, then some person down the chain of command will likely pass the coordinates on to the surveyor as the place to put the well. Coordinates are what a surveyor deals in, and it is natural to stake the coordinates as given and to wonder why the shot point isn't there also. It must be made clear that the coordinates are only to help find the shot point, that the stake is to be put on that shot point, and that the coordinates of the shot point are then to be surveyed as accurately as possible.

Offshore, the problem is more complicated. There aren't any tags, bench marks, piles of cuttings. The lines were shot using radio surveying, and the energy source was an array of air guns or something similar that doesn't leave evidence behind. So it is even more natural for the well recommendation to be in coordinates and for an effort to be made to locate the well exactly on those coordinates. But, in surveying terms, the key word here is "repeatability." Here too, the shot point is the place to put the well. The most accurate way to do that is to use the same type of surveying the crew used. Most offshore crews use a system that measures radio signals from fixed stations on land. In the effort to go back to a shot point, the same type of equipment and, if practical, even the very same pieces of equipment—with the same land stations, antenna heights, etc.—should be used.

After the rig is positioned, then it is advisable to use several days of satellite readings, or whatever seems most accurate, to see exactly where the location turns out to be. This information is useful if the well is a discovery, and the rig goes on to other locations, with a platform to be erected later at the drilled location. The accurate location is also necessary for future use in plotting the well on maps, satisfying government requirements, and using the well as a fixed location for surveying in the area.

WAIT ON THE DRILL

Now comes one of the hard parts of your interpreted prospect, proposed well, hoped-for oil field, maybe secretly dreamed of giant field—waiting and bearing up.

If management decides to drill the prospect and money is set aside for it in the budget, then some months may pass before there is a rig available and assigned to that well. There are delays in waiting for the rig to be near that location so money can be saved by making a short move, waiting for the site to be made ready, waiting on a suddenly urgent other prospect to be drilled out of its normal place in the sequence, delays for all sorts of reasons. Then the rig is on the site, and if you are in a small enough office, you may get to see the daily reports as they come in— rigging up, spudding, setting surface casing, drilling cautiously through a tricky zone. You wonder why they won't get on with it and get down to the pay zone, the formation that is expected to have the hydrocarbons.

Then the well is nearing your zone, and the geologists start asking you why it's running low, encountering formations deeper than expected relative to other wells. My gosh! Is it possible that it can be dry after all? Hang in there. Just wait, you don't really know yet. The bit isn't into the formation yet. Then it reaches the formation but you still don't know. You have to wait till the logs are run. The logs will give fairly good evidence, but if the formation looks good on the logs, it may be necessary to wait for drillstem tests, opening up the hole and seeing what comes out of it. Then you can know. In the heat of the moment, people will be inclined to give you more credit or blame, as the case may be, than you really deserve. Accept it, without trying to explain too much. Whichever you got too much of, you may get too much of the other next time.

If you are given credit for a highly accurate prediction of the depth of a formation, don't admit to that much accuracy. Suppose you decided, for the well prognosis form that is filled in before drilling, that the important formation would be 3,050 meters and it turned out to be 3,051 meters. That was chance. Seismic data isn't accurate to one meter at that depth. If people think your prediction was really that accurate, they'll want your next one to be as good. It isn't going to be, and there could be real blame if they thought you should be able to do that well. Explain that you'd have been glad to be within ten or twenty meters, and only luck put it just one meter off.

Well, that's over. The waiting is done, and there is an oil or gas field to develop or, more likely, a dry hole to figure out and try to profit from.

AFTER DRILLING, WHAT NEXT?

Either way a well turns out, there are bound to be some surprises. We just don't know that much about the subsurface from seismic investigations conducted up here on top of the ground. The geologists will be talking about an extra sand that was found in the well. You will wonder why the well produced even though it was low or wonder why it was high, but dry. The logs taken in the new well give you a whole set of new information to be fitted in with the seismic sections, other wells, etc.

USE THE KNOWLEDGE

If the well was a disappointment, there may be some pressure to explain why it was dry. No one is really happy at the outcome, although experienced oil people expect a lot of that kind of disappointment and tend to take it in stride. You certainly need to incorporate what you have learned into the picture you have of an area. It may be that some of the new information will lead to a discovery somewhere else in the area.

If the well was a discovery, everybody is happier but the analysis of the new information is just as necessary. There is a new field to develop, wells are expensive, and they need to be positioned as economically as possible.

IF YOU HAVE A DRY HOLE

When an exploration well is drilled, the most likely outcome is that it is dry—not no-water dry, but no-oil "dry," of course. Even though that was the most likely outcome, it was probably not the expected one. When you work up a prospect, you get to believing it. And when people approve money for a well, they get their hopes up. Although everybody concerned knows that prospects usually turn out to be duds, they find it hard to give up the hope that this one would be one of the good ones. Well, it wasn't. It's time to think about two things, why it was dry and how the information learned from it can be used to find oil.

First then, why was it dry? You had thought up the reasons why it might produce, and beforehand you didn't want to stress the negative side, or prospects wouldn't be drilled. But now the negative side has prevailed, and you need to determine, if you can, which of the possible negative factors took effect, either among those you had anticipated or some other one you hadn't thought of. This rehashing is not being pushed as a punishment to you; it is not a thing to begrudge, but rather it is an essential part of the company's exploration program. This first question, why it was dry, is part of the second question, what can be learned from the well that may help in finding oil.

To answer both questions, you should work closely with the geologists. They will inspect the logs from the well, correlate them with logs from other wells in the neighborhood, learn all they can from the well. The well was probably drilled on a seismic line. This situation now is one of the reasons why it is useful to drill wells on lines. If the well is not on a line, then you can't now draw very reliable conclusions from it, about its relationships to the seismic data. Any relationship that looks strange may then be because there is a variation in the geology somewhere between the well and the line. But if the well is on the line, then well data and seismic data can be assumed to be from about the same place, so you can feel that they must be made to agree.

There are some reasons, though, that can cause the well and line to not be at the same spot, even though the well was intended to be drilled on the line. Either could be mispositioned by a surveying problem. This isn't likely if the line is fairly new. There are location problems caused by dipping underground layers. A drill bit tends to work its way in the updip direction, so information on the deviation of the well should be checked. Also, and fortunately in the same direction, seismic data comes from a position updip from the shot point. If the section has been migrated, that takes care of the displacement in the dimension along the line, but not in the direction to the side of the line. In dipping beds, a seismic line that is not exactly aligned in the direction of dip produces a section that represents data from one side of the line, the updip side. With depth, the displacement increases, so the section represents a tilted more or less plane. The "plane" is irregular, as dips and shapes of beds vary. If a well is drilled on the surface position of the line and there is appreciable crossdip, then at the depth of the formation of interest, the well, if drilled vertically, will not be at the depth point seen on the section.

A way around this position problem is to select the well location before drilling, with these factors in mind. Talk to the drilling engineers and the geologists about how much the well is likely to be deviated or could economically be allowed to deviate. Work that out with the cross-dip as determined seismically from other lines and your map, determining the offset of the well that the dip will cause at the important depth. You might get a location and a tolerance of deviation that will both save money and effort in drilling and produce a better match of well logs and sections, so the well and section coincide at the target depth.

A wildcat well based on seismic information should have a sonic log run from the bottom of surface casing to total depth and a velocity survey with check shots. Then a synthetic seismogram should be made from them. A vertical seismic profile can also help. All this is to enable you to relate the well to the section accurately enough to draw valid conclusions from them.

When the well has just been drilled, the information won't all be in at once. Logs, synthetic, etc., will come along as they are made. If there is an urgency to getting the problems analyzed, you may well be working on partial data in rough form. That's the way things are. The next location for the rig, or some other operation, may depend on the information. The whole company can't stop while you wait for all the data, so you can do your work more efficiently. If there is time, fine, it may make sense to wait. Even then some time spent with the preliminary data may help you to know what is going on and to be able to answer some questions when they come up.

What you do with the new data is very like making the original interpretation. First, you need to identify the horizons again, using the data from the new well. This identification may need to be more precise than the first one, as you are answering more specific questions. Then if things don't happen to fit, you may need to go over your picks in the vicinity of the well. It may tell you that some of the decisions you had made in picking must be changed. After the identification and picks become consistent with the well, the map may need to be contoured again.

The new map probably isn't one that will lead to an immediate prospect, but it can help you understand the area and it may lead to other prospects. You have learned a bit more about interpreting the area and have a mapped feature that you know about from drilling. This feature can be compared with others in the area.

The drilled prospect may lead you to decide that more shooting on it or on similar features would be useful. Or it may set you off on another interpretation of the area in the light of the new knowledge. Sometimes a dry hole may turn out to be a very good thing in terms of eventual discoveries.

IF YOU FOUND OIL

If the well found oil, enough oil to be economically worth producing, then there is a similar need to rework the data. A discovery has been made, and everybody is happy. The pressure is off. But in the immortal words attributed to Harry Sinclair, "Don't sit around eating chocolate candy; get to work and find some more oil." The new information is just as new, and the seismic data and well data need to be made to agree just as much as if the well had been dry. There isn't the question of why it didn't produce, but the well probably contained some surprises, so there may be a question as to why it did produce. That's shocking. You turned out to be right in that it did produce, didn't you? Why the doubt? The well that found "your" oil may have shown your interpretation to be less than completely correct. And now the interpretation matters even more. There is the same need to learn about the area so other fields can be found, but there is also a field to be developed. The seismic mapping will be necessary to help locate delineation wells that will find the limits of the field and also locate the development wells needed to drain the field. It had better be as correct as you can possibly make it.

When a field has been discovered, one thing is certain—a great deal of money will be spent on it. If it is on land, rigs will be set up and wells drilled at enough locations to drain the field, and some will be drilled outside its limits, to establish where those limits are. The drilling of each of these wells is expensive, and they will be maintained and occasionally worked over through the years. Gathering lines will be put in to connect them with tanks or pipeline. If they are too far apart, some of the oil or gas that might have been produced will stay in the ground. If they are too close together, there will be more expense, etc., than necessary.

If the field is offshore, the situation is similar but far more costly. Platforms costing tens of millions of dollars each will be set up at carefully calculated locations, and several directional wells will be drilled from each platform. The losses from spacing the platforms too close together are obvious, with platforms so costly. If they are too far apart, there isn't, economically, even the stop-gap procedure possible on land of drilling more wells between the original ones.

On land or sea, it is very important to have all the information possible before positioning development wells. A first step is to rework the data that led to the discovery in the first place in the light of the new well data.

If time permits, you can profitably use the logs, synthetic seismogram, whatever you have from the well, to make a totally new interpretation of the prospect. First, you can identify the horizons again on the section from this well in the midst of the prospect. Then you can pick the sections again very carefully, with a different attitude. In exploration, you try to find something to drill. Now you have the field and are more concerned with detail and certainty. Anything on your map may now be subjected to specific, even picky, examination by geologists, engineers, managers.

With all the money that is at stake, the investigation should not stop there. You really need more shooting.

PART D

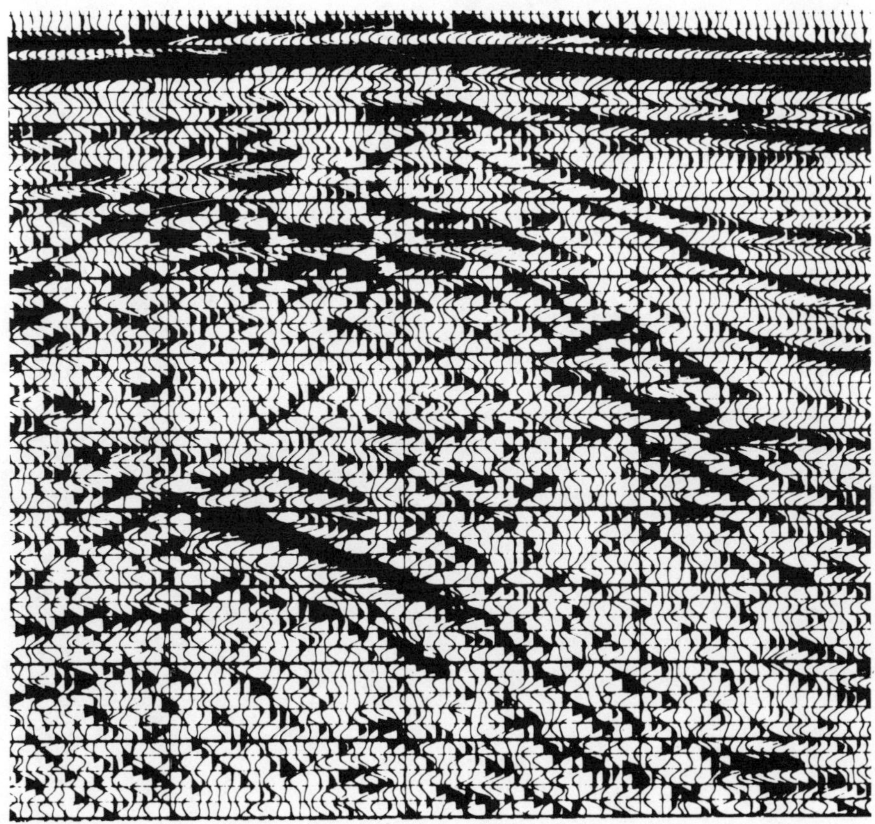

ET
CETERA

Plan a Program

Determining a seismic program to be shot requires considerable seismic judgment. Program planning obviously is a step that normally comes before interpretation, but for this book it seemed better to start with interpretation itself. New interpreters usually begin with areas that have already had the program planned, shot, processed, and sections delivered. Sometimes the programs are planned by people who are not seismic interpreters. But planning programs is probably best done by interpreters, and well-laid-out programs are needed for the interpretation to be effective. Now we have got to a point in the book where a detail program is necessary to the development of a field. Let's let the field wait awhile so we can take up ordinary exploration programs first and work up to the detailed programs needed for development.

EXPLORATION PROGRAM

To start exploring a new area, people usually shoot some kind of grid of lines over it. If absolutely nothing is known about the geology of the area and if it is all equally easy to get around in, then a square grid can be shot. Parallel lines in one direction can be crossed by another set of parallel lines at right angles to the first set. The lines can be spaced far apart, with the distance dictated by financial and timing requirements. That is, the less money available for the shooting, the farther apart the lines; the greater the hurry to develop the area, the closer together they can be. This type of program and some others we'll discuss are shown in Illustration 12–1.

Let's add a few complications to that simple situation. The condition of knowing nothing at all about the geology is rather unlikely. There is usually some knowledge to go on. If there is none, a company is unlikely to want to spend money on the area. Suppose some regional geology is known, enough to show the general grain of the structure in the area. Seismic lines that extend up and down dip are the most useful. Those along strike tend to show little structure and to find little oil. But the strike lines are useful to tie the dip lines together and keep the interpretation from going astray. You don't need strike lines as much as dip lines, but you do need some of them. The grid can be rectangular with dip lines closer together than strike lines. This is the way most people assign reconnaissance lines, with a grid of three by four kilometers or something like that.

Now let's add another complication by considering an area with obstacles to getting around or shooting. The obstacles can be physical or legal, natural or artificial, onshore or offshore, year-round or seasonal. There can be mountains, rivers, towns, no-permit areas, power lines, swamps, shipping lanes, anchorages, fishing nets, mine fields, roads, railroads, political upheavals, angry landowners, lambing seasons, freeze-ups, thaws, unharvested crops. I'm sure you can add to the list. In spite of all this, lines tend to be assigned from nice white maps, just wherever it looks like the lines might be useful.

Sometimes this is necessary. The field crew can usually somehow manage to go almost anywhere a line is assigned. But it may be costly in money, time, data quality. It is best to have topographic maps, air photos, satellite photos, political maps, land-use maps, whatever might be helpful. It may be that a line assigned to break up a loop into two smaller ones crosses a river. Perhaps a line in another direction, parallel to the river, would break the loop just as well. That line would probably cost less to shoot and might have better quality data, too.

For common-depth-point shooting, lines need to be straight or nearly straight. But maybe a line can be moved over or a piece of it skipped. It is better to plan these changes than to have them confront you after the shooting has started. With foresight, the changes can often

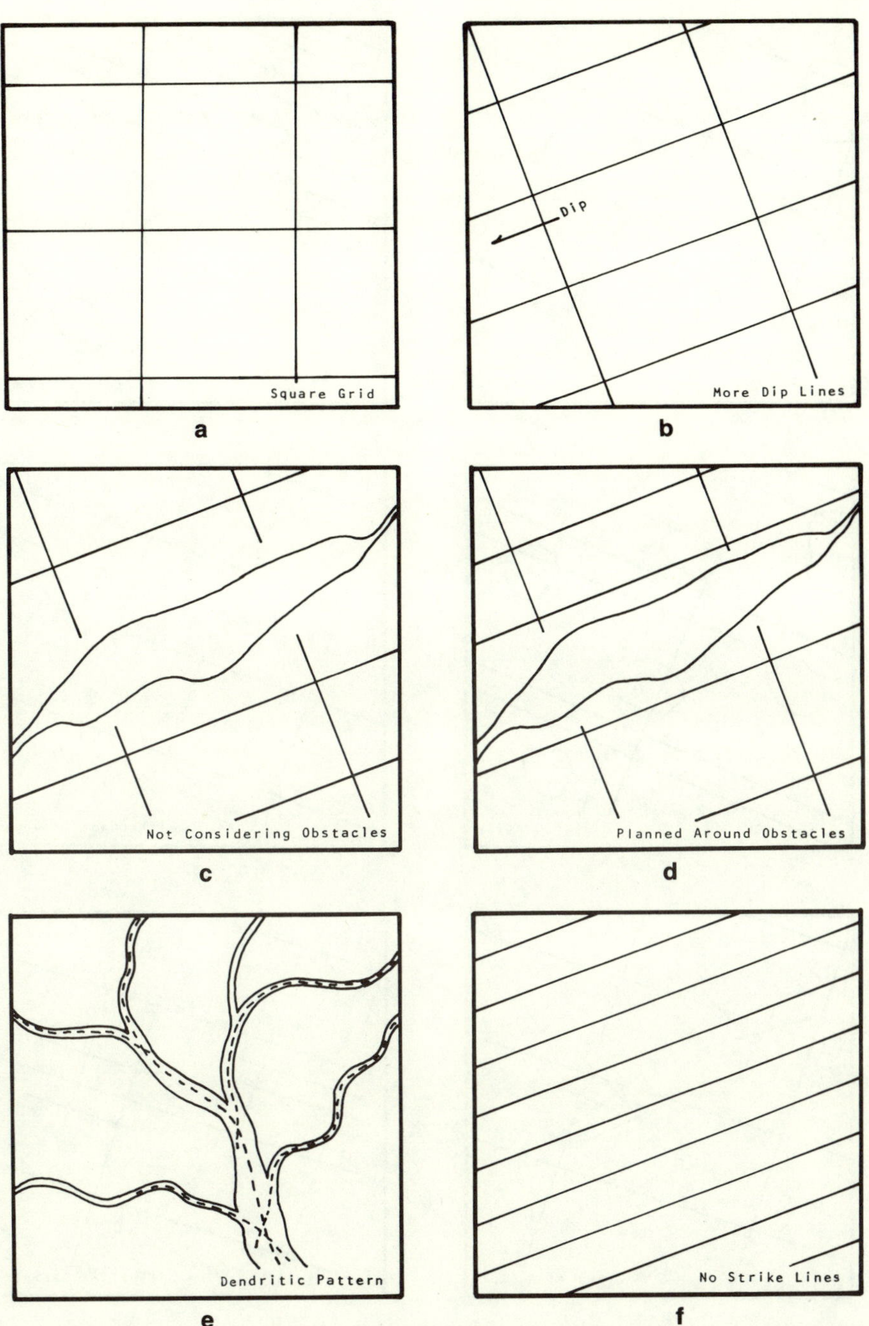

Illustration 12–1 Program planning

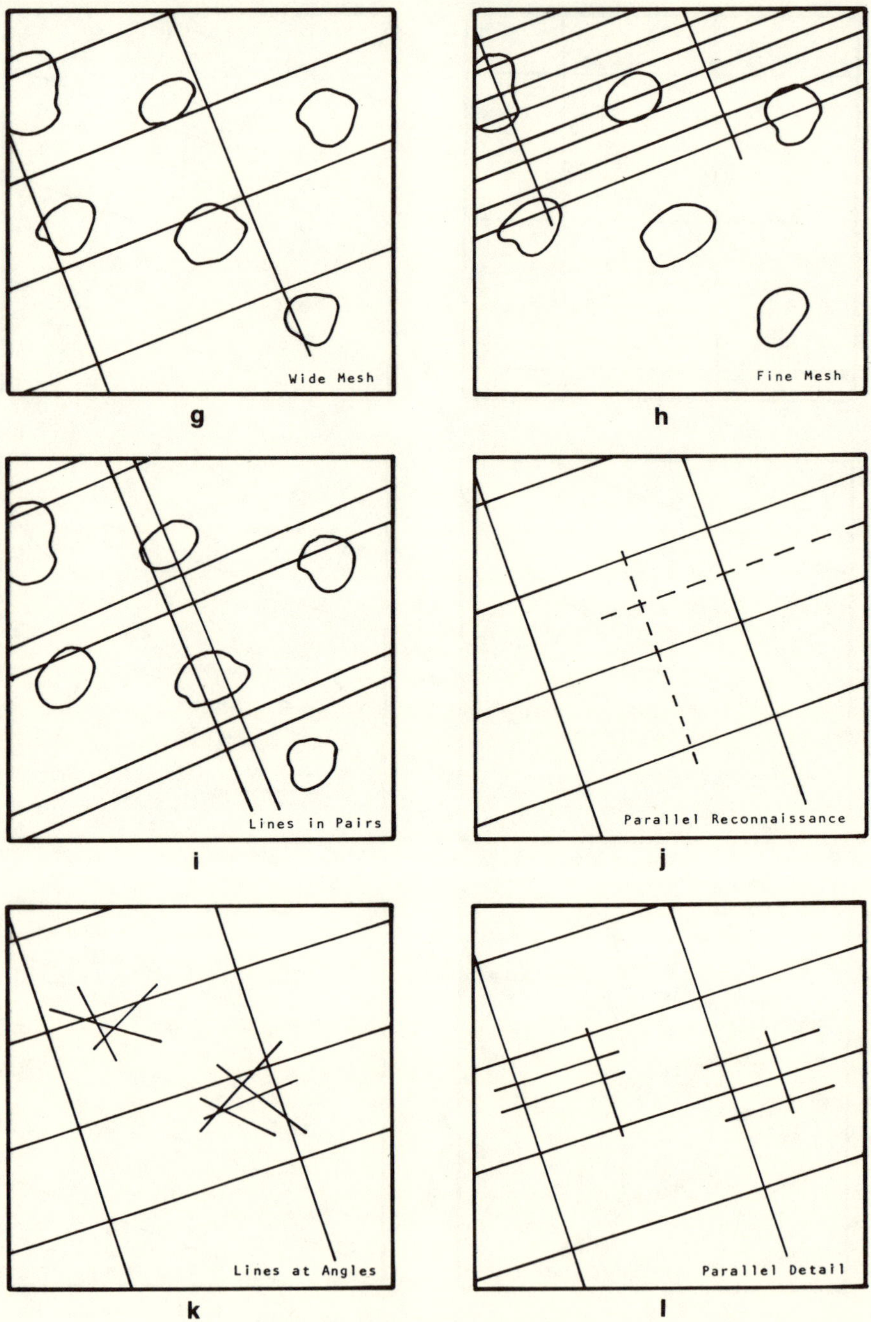

Illustration 12–1 (continued)

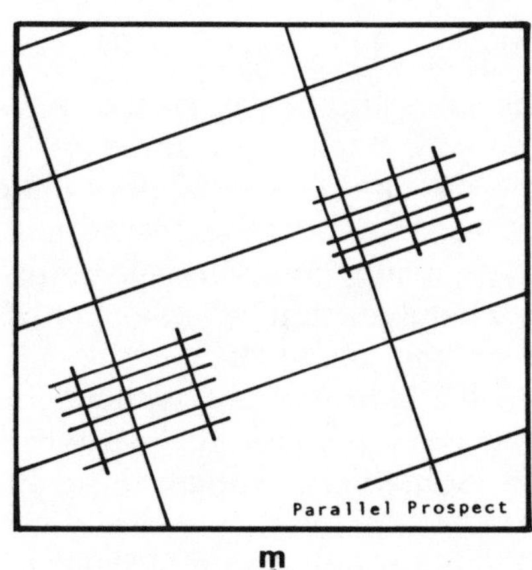

m

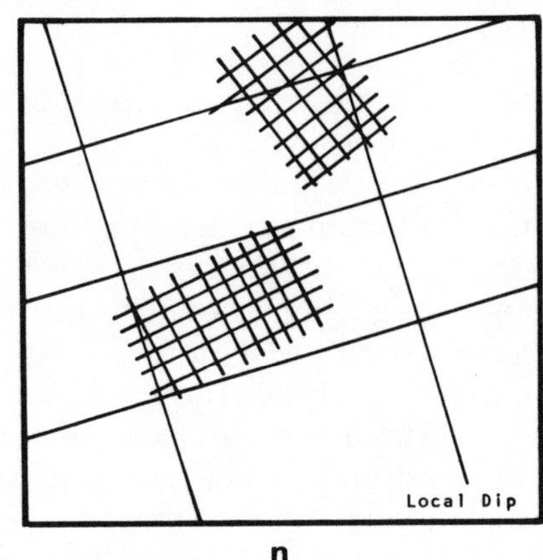

n

Illustration 12–1 (continued)

be less harmful than running into a problem halfway down a line. Even with planning, there will still be problems in the field, but some of them will have been prevented.

In a new area of difficult access, the use of ingenuity may lead to reasonable coverage in some way other than just forging ahead. A land area with jungle or steep hills may lend itself to shooting from a boat in the rivers of the area. This might cost a tenth as much and take a tenth of the time of a conventional straight-line land survey. The boat survey won't have straight lines for a high degree of stack to improve the data quality, but the absence of an aerated zone near the surface may make the river records better than the more highly stacked ones on land. Now that kind of coverage is well worth having. Or there might be a pattern of ridges that can easily be shot in swampy country or a pattern of valleys in rugged terrain.

Interpreters usually need to have their lines arranged to form loops, so they can check to see if they have tied correctly when they come back around to the starting place. If the reflection quality in the area is poor, you may need many small loops to keep you on track. With better data quality, loops can be larger and can have a higher proportion of dip lines to strike lines. With still better data, the strike lines can be still farther apart. In the extreme case, reflection quality might be so good, and recognizing a horizon so easy, that you might not need any strike lines at all. Then you could just correlate from dip line to dip line.

The first requirement, then, is that lines must be spaced closely enough to satisfy the local reflection quality requirements. After that, the spacing of the dip lines is dictated by the size of prospect being sought. If you are just looking for big structures or regional stratigraphic traps, the lines can be quite far apart, within the limits imposed by data quality. If the features you are looking for are small, the only way to avoid missing any of the features is to have the dip lines closer together than the width of the smallest feature. Whether it is worth the cost to find every feature in the area is a decision to be made after you have determined, or guessed, how much shooting it would take to find them all. If you don't want to hunt for all of them, then you can decide whether it is better to cover the whole area with a loose grid or part of it with a tight grid. Without really thinking about the decision, people usually use the loose grid and maybe gradually fill it in. Sometimes though, the tight grid, later extended further, may make development of an area of small features easier. It allows drilling, gathering (pipe)lines, other facilities to be handled more economically. With either plan, a feature is only found, not outlined, by one line across it. Every feature found will need detail to map it and find the most drillable spot on it.

A later phase of reconnaissance shooting should in general be made to fit the grid already established, even if it has been found to not quite agree with regional dip. Lines parallel to the existing lines will cut the area into smaller rectangles. A different alignment for the new grid would make for a pattern of odd-shaped large and small spaces between lines. These large and small spaces would find some features and not others, like making a fish net smaller by adding more string across just some of the holes.

PAIRS OF LINES

To get some of the advantages of 3-D, but with 2-D shooting, just program pairs of lines close together. In plotting seismic programs, people tend to design grids with lines equally spaced. But if instead the lines are in closely spaced pairs and the pairs are farther apart, then a pair of lines sufficiently close together will give directions of faults and dip directions of features. There is still a grid, but of a wider mesh than there would otherwise be.

A lone seismic line crossed by a fault does not give any idea of the direction of the fault trace. But if there is another line nearby, with a fault on it near the first, then it may be easy to relate these two fault indications as belonging to the same fault. A line on the map drawn through the two fault indications shows the direction of the fault trace.

Similarly, if a line crosses an anticline, there is no indication on the section of whether the line happened to be on the crest of the structure or on one flank or the other. But on two close lines, the structure is likely to be higher on one than the other. This shows which way is updip on the feature and, therefore, which side to do further detail shooting on.

Arranging the lines in pairs need not cause any extra expense. As pointed out at the start of this topic, larger gaps can be left between the pairs. This means that if the arrangement planned was a uniform grid tight enough to not let any prospects be missed, then the wider spacing will no longer accomplish that. However, it is probably better to know more about some prospects than to merely detect all of them. As already described, omitting some of the strike lines from a rectangular grid can perhaps keep the grid tight enough without excessive expense.

DETAIL A PROSPECT

After reconnaissance work has produced some prospects, there is reason to do some detail shooting on them to verify that they are prospects, to define their shapes and sizes, and to find the best places to drill on them.

There are several ways to shoot detail lines. The traditionally most common way, and probably the poorest, is to try too hard to save money by assigning one or two carefully planned lines across the feature. But you probably don't yet know enough about it to decide just which lines will give you the information you need. The line or so rarely provides the answers wanted. The same prospect might have to wait for more detail in the next seismic program, maybe a year later. Then another line or two may be shot. And so on. Also, these very carefully planned lines are usually at various angles that happened to look best on the prospect as mapped at the time, so the result is often a complex of crossing lines, laid out like the sticks in a campfire.

There are more effective plans, using one form or other of grid. A feature may not really be a prospect yet but merely a lead, a hint of a prospect. It may seem to need only a line or two. Or the company's budget may only allow a little shooting at the time. Then one or two lines parallel to the lines of the reconnaissance grid will get more information per kilometer of line in terms of improvement in the contouring of the prospect. These parallel lines will divide the control for the prospect into smaller rectangles, rather than a mix of large and small triangles and other shapes.

A prospect, more developed than a lead, can be handled more effectively with a close grid of lines across it. The reason for all the cost of shooting the reconnaissance lines was to find prospects. This is a prospect, so it is worth a more thorough investigation. A close grid in both directions is a good idea at this stage. One way to shoot this grid is to make it fit the reconnaissance grid. This fits the existing pattern so that old and new lines can be used effectively together.

Another plan that some companies use is to determine the general dip across just that prospect, from the existing data, and then plan a grid oriented to that dip. After that grid is shot, they make a new interpretation of the prospect, using only the new lines. This costs a little more, as the new grid must be complete, not just filling in between other lines, but it is a good way to directly look for oil on the prospect.

In Exercise 12–1, plan a reconnaissance program and a detailed program.

Now we've covered planning programs for exploration. As promised, we'll get back to where we left off and discuss the type of shooting necessary to develop an already discovered field, exploitation shooting.

DETAIL A DISCOVERY

The shooting to delineate a discovered field cannot be postponed or done as needed from time to time. Once the installations are in place, the detailed seismic control that might be needed cannot be shot. For one thing, the rigs, pipes, mud pits, buoys, anchors, tankers are in the way of shooting, so spreads cannot be laid out where they are needed. Also, the sounds of engines, cranes, generators interfere with the recording of seismic data—and if you think it is easy to get everything shut down for a seismic crew, you try it! The people on the rigs feel that, with all their big equipment, their work is the important part of the operation, rather than some little seismic crew.

Therefore, the shooting has to be done right after the discovery is made, that is, before any equipment is moved in to develop the field. And the shooting has to be as complete and as up to date as you can get. There won't be a new program next year, not a good one anyway. A line or so might be squeezed in somewhere at extra expense and not shot in just the place and manner needed. The program for now must be complete, with lines as close together as you think you may want. You need detailed information about sands that come and go, or about irregularities of the surface of a reef, and whatever the engineers, production people, and exploitation geologists can think to wonder about, that might be resolved by geophysics. And the shooting can't include next year's innovations in cable and recording. So what you do use should be the best, most modern, most versatile that is available at present.

It is just the shooting that can't be re-done next year. New processing can be done, new techniques used on the data, whatever is new or that the company is willing to pay for at a later date. The only new processing techniques that won't be available to you later are those that require the use of a new technique in shooting.

Your recording should be done with a sampling interval of one or two milliseconds, not four. You should record a large number of traces, with a large CDP multiplicity. The spread arrangement should be primarily designed for optimum recording at the depth of main interest in the field. But if there is extra capacity available in cable and geophones, it can be used to permit the investigations, which will surely be requested, of other zones encountered in the wells. For that, if there are enough extra channels, etc., the spread should be long enough to record deep information and also dense enough at short distances to give good shallow data.

Notice that I'm now acting almost as though we should ask management for all the money we want for shooting. What happened to frugality? Well, this is a different situation from exploration. In exploration, we are hoping to find hydrocarbons. Every bit of exploration expense that does not find anything is, in a sense, wasted. Obviously we have to do the exploration. If we knew in advance which part would find prospects, we wouldn't need to shoot the rest. But we don't. That's what exploring is all about. Now, though, we aren't exploring. That's been done for this field. The field has been found. Now a lot of money is going to be spent on it, and we have a chance to keep a lot of it from being wasted. We can do that by spending an amount that would seem fantastic in exploration but is small in comparison with these other expenses. The optimum placing of a forty million dollar platform will quite easily justify a seismic expense of a few hundreds of thousands of dollars.

For this program, then, you need the lines oriented along dip and along strike. It's fine to have them close together in both directions now. And if the lines you plan are so close together that it almost amounts to a 3-D survey, then it is probably better to go ahead and shoot 3-D. You can do more with it later on in processing.

And make the lines long enough. In developing a field, there is a temptation to shoot lines that just barely reach across the field. But some of the questions that will come up later will call for knowledge of the conditions off the field. There might even be a new idea that could lead to extending the field. Of course, the lines need to be long enough to function well as seismic lines. The ends must extend far enough to give full stack as far as you need data and enough extra so that, when the line is migrated, there will be migrated data as far as needed. The ends of migrated sections, not having data beyond them to migrate into the ends, are not fully migrated but are a mass of smiles, the same up-curving events that are seen at the bottom of migrated sections. This distance depends on the steepness of dip in the area. Fairly flat reflections do not migrate far; steeper ones move farther. In a steep-dip area, you may need an extra kilometer or so at each end of the line for effective migration.

All this justification of seismic effort that I've been talking about is somewhat dependent on the size of the discovered field. One that looks like it will be a large field will have more money spent on it, so it can more easily be shown that the seismic expense will save or earn more money. In one way, though, a marginally commercial field might need detailed shooting even more than a big field, to better ensure that it will be commercial. Even its development expenses will be large in comparison with the seismic expense that could guide that development to profitability.

In all the discussion of exploitation shooting, I have been somewhat indefinite. Necessarily. I don't have your specific discovery to look at. You will have to think about the situation and see what the problems are to decide what shooting program will do the most for it.

CONTRACTS

Contracts for shooting and processing are not a part of interpretation. They will normally not be a concern of the beginning interpreter in an organization with a large geophysical department. But geophysicists work where there is opportunity. You may go to work for a small company, or even in a large company, you may be stationed in an outlying office. In either case, you may be the only geophysicist in the office. If you are, you are likely to find yourself the nearest thing there is to an expert on any or all of the aspects of geophysics. If the decision is made to hire a contractor to do some shooting, you may well be asked whether a certain kind of contract should be accepted.

With this in mind, let's go over some of the main aspects of geophysical contracts. A major factor is price. We're fairly safe on that. By asking for bids from several contractors, you find out what the going rate is. If you don't want to go into a formal bidding, you can just ask around among the contractors.

In a large organization, bids are usually required. There are some problems associated with bids. They tend to lead to a necessity to accept the lowest bid. The lowest, of course, may not be the best. Higher quality may find more oil for less money than the cheapest shooting. Also, the examination of bids is a difficult and time-consuming process.

There are two basic methods of calculating cost in seismic operations, by amount of shooting or by expense. Neither is commonly used in pure form.

The first, turnkey work, is charged for by the shot, profile, kilometer—some unit of the shooting done. The second is charged by the expenses incurred by the contractors, with enough added to those costs to allow them to earn some money.

Turnkey work is convenient. A budget can be made for the exact amount the shooting is to cost. The bill is simple, so many kilometers at so much per kilometer. There isn't much checking necessary to see that that many kilometers were actually shot—you have the sections to show that it was done. And turnkey bids are easy to compare.

There are disadvantages to turnkey work, for both oil company and contractor. The turnkey price has to be high enough to allow the contractor to not go broke if there are a lot of obstacles to getting the work done, bad weather, rough terrain, permit difficulties. The main incentive to the contractor is to shoot as fast as possible. Contractors want to keep their reputations, but they do lose money with every delay.

The other type of contract, with billing based on the contractor's expenses, gets around the disadvantages of turnkey work. The oil company is hiring the crew to do whatever is needed. Time spent in experimentation or in a slow and careful shooting method does not lose money for the contractor. The oil company can have the work done as thoroughly as it is willing to pay for.

There are disadvantages to this cost-plus way of billing. The billing is very complicated, with the need to account for each item of expense. It is to the contractor's advantage to have the work take longer and cost more. Again, contractors are mostly ethical, but they can't totally ignore their own advantage.

Somewhere between the two is a billing on a time basis, so much per month or per day. This is related to the contractor's cost and gives protection from downtime because of weather and other problems. It encourages the contractor to keep costs down.

The disadvantages of the first two methods are severe enough to make them not usually practical for seismic contracts. A blend of mostly time and cost plus is normally used.

The contractor's basic personnel and equipment are usually billed on a time basis, some flat fee per day worked. Equipment beyond the standard—like extra drills—and additional personnel—like extra surveyors—are billed as additional time items. Some expendables, like explosives and food, are charged on a cost-plus basis. Thus, the contract is flexible enough to keep the oil company's and the contractor's goals both satisfied fairly well, without undue hardship on either one. An additional charge per kilometer shot is a turnkey element that may be used to mix the incentives up a bit.

This type of contract makes comparing the bids of different contractors quite difficult. One will have an item included in the basic crew that another charges extra for. One may include more drillstem in the charge for a drill. One way to compare the bids is to assume some arbitrary amount of time, personnel, equipment, supplies that you think is likely to be needed when the work is done and calculate what each contractor would charge for that. This isn't really a fair comparison, as your assumption won't turn out to be correct when the shooting is done. But to compare fully would require a large number of similar assumptions, far more than a person can handle in a reasonable length of time. Maybe someday someone will write a computer program to make all the comparisons.

Data processing is not so dependent on other factors, like weather, so a contract for it is likely to be closer to the turnkey type but partly broken down into separate items. There is usually a decision as to what is to be the standard processing for the data. This will be contracted for on a basis of so much per profile, or kilometer, or other unit. Then additional prices will be set for extra processing on a similar unit basis. The standard processing will include a certain amount of experimentation and a certain number of velocity analyses that are to be done without extra charge.

You plan seismic programs in a consistent way, so new shooting adds as much useful information to the old as possible. You take the special conditions in the field into account. And work out the contract to give your company what it needs for a good price, without hurting the contractor so much that corners are cut in your shooting and processing.

Problems, Problems, Problems

This chapter is not your introduction to the problems of seismic interpretation. Problems are all through the book. These, though, are some that might have been neglected without a special place for them. They involve the questions of how good seismic data is and how good the survey data is.

RELIABILITY OF SEISMIC DATA

After all the expense and effort of shooting a seismic survey, processing the data, and interpreting it, what do you have? Usually a map and some impressions about prospects on it. How much can you trust all that? Was it worth doing? You hope it will improve the chances of finding oil. We know it won't make it certain. Many dry holes are drilled on seismic recommendations every year. But if seismic investigation makes the odds a little more in favor of finding oil, that will make it worth doing.

Wells are so expensive that we don't want to cause dry holes to be drilled. On the other hand a new field can be so profitable that we can't take the cautious approach of drilling only sure prospects. There aren't ever any sure ones. If people waited for them, no wells would be drilled and no oil discovered—none at all. Deciding what prospects to drill is a matter of judgment and, in the end, of guessing about the risks and the chances for profit.

PROBABLE ERROR

Back to the original question—how much can we rely on seismic results? Would it be worthwhile to put after each seismic time on maps a probable error, like plus or minus four milliseconds? That notation would be nice and satisfying. You could feel that people had been warned so they could take proper action. But there are two reasons why that isn't practical.

First, the probable error doesn't mean much. It could be applied to the slight inaccuracy of picking times and the similar error caused by not quite adequate near-surface corrections. But what about having jumped a leg in picking and being 50 ms off? Or being a leg off and then having the leg you're on and the correct one diverge, so you're 200 ms off? Or having misinterpreted a fault throw and being 500 ms off? All these things can, and frequently do, happen to interpreters—new, inexperienced interpreters and highly skilled old hands at interpreting.

The second reason why the indicating of possible error isn't practical is that there isn't much people can do with the information. Other geophysicists might understand it, but they already know that the times aren't exact. They don't need a guess at every shot point, or even an overall guess in the form of a note in the legend of the map to remind them.

Also, the people who look at the maps to decide whether to drill a well include geologists and managers. These people probably don't know what the fine points about the data mean and can't be concerned with them in making decisions. The seismic map is usually just one of several things to be considered in deciding whether and where to drill a well, or whether to lease an area, or whether to shoot another seismic survey. And what these people must look at is the overall map, the significant features on it, and the confidence the geophysicists have in it. Details about data don't matter in the critical use of the map, but they do clutter it up with unnecessary numbers. Some companies even go so far as to leave all data off their final maps. This is extreme, though, and often in meetings would make it necessary to dig out the work maps to see how much the contours are in accord with the data or whether the data could be contoured another way.

WHY DIDN'T IT PREDICT THE DEPTH?

When a location is recommended for drilling, the managers naturally want to know how much it will cost to drill the well, the production people need to know how much drillstem to order, and so on. A lot depends on how deep the well is to be. The geophysicist is asked for the depth to the pay zone and to the planned total depth. Using the velocity data available, you estimate the depths the best you can. The well is drilled and the formations are found, not where you said they would be, but somewhere shallower or deeper. People wonder why you couldn't have given them correct depths. You try to explain about the vagaries of identification and of velocity determinations, but they don't really want to know a lot about the problems of a geophysicist. They have their own problems. They just think you should have done a better job of estimating.

If the formations are deeper than predicted, the well will cost more than planned and money will have to be obtained by pulling it from some other project or something. If there was not enough drillstem or mud or something on the rig, there may be expensive rig standby time run up while more is obtained. A too-deep estimate can also cause problems, with the drill reaching a layer before it was expected. These problems are usually not as bad as the other problems, but they can still be severe.

Make the estimates with some slack, by adding to the depths or by saying a formation will be encountered somewhere between two depths. This kind of expectation that we can't fulfill occurs frequently because people in other fields don't understand what things we do better and what we can't do so well. Let's go over the things seismic interpretation does. Maybe sometimes you can improve communications with geologists and managers by telling them in advance what can be expected of seismic data.

Depth to a formation is what people unfamiliar with geophysics would most expect seismic work to yield. It's what comes from well data. But it is one of the things seismograph is poorest at, for two reasons. First, a seismic reflection is not recognizable as being from a certain formation, like a rock sample might be. It's just a reflection. In general, it is identified at a well that has a velocity survey and is then picked along seismic sections to the location of the proposed well. The identification at the velocity survey is usually good, but the picking may go astray somewhere along the way, maybe even far astray if some fault or other change is interpreted incorrectly.

Second, even if the identification is good, its depth isn't known at the proposed location. The reflection is on the section at a seismic time, which must be converted to depth with some velocity information. Velocities vary from place to place and usually cannot be very precisely determined if there is not a well nearby. And most prospects are not where there is a well, but where one may later be drilled.

Having the processors convert a section to depth is not a precise solution to this problem. A depth section may look more like the true geology than a time section, but if you plot formation tops from well data on the depth section, you are likely to be in for a rude shock.

People may also ask what kind of lithology is represented by a certain reflection. We can't tell. A reflection is a reflection. It comes from an interface between rock layers that transmit sound at different velocities. That can be almost any two kinds of rock. There are some clues in seismic lithology, but they are usually not definite enough to allow you to point to a reflection and say, "This is a sandstone." If the reflection is correctly identified, we may be able to say what formation it represents. That in turn will indicate what the rock is made of, if there hasn't been a lithologic change between the point where the identification was made and where you are now.

We don't know depths very well, so we can't tell about exact dips, either. You may see a dip on a section and obtain from it an estimate of the depths of the two ends of the segment. Any lateral velocity variation above that horizon will distort the dip, making one end deeper than it should be in relation to the other.

The direction of dip is usually fairly reliable on a seismic section. Velocity variations can make a reflection appear to dip in the opposite direction from its real dip, but this is not common. Rock layers generally dip in the directions that they seem to dip on seismic sections.

One of the most useful aspects of seismic sections is that the reflections on them are continuous, rather than being isolated like wells.

SEISMIC DATA VS. WELL DATA

Seismic data are often compared unfavorably with well data and other information. But it should also be pointed out that some of the strong points of the seismic method are in areas in which other approaches to geological truth have their shortcomings.

Well data have traditionally been thought of as being "right." The attitude has been that the well was drilled down into the earth, so there can be no question that its data, obtained from some instrument's actually being down in the hole, is superior to seismic information, which must depend on receiving echoes from underground. That has been the position not only among geologists but among geophysicists too. If a seismic section and a well do not agree, then the section, or the interpretation of the section, must be made to fit the well data.

However, the well data has a big disadvantage, too. The wells are far apart. Correlation from well to well is subject to the usual problems of correlation. The correlation might not be right. And even a correct correlation does not tell where a change takes place between the two isolated points. The change might be anywhere between the two points. A thinner zone in one well than in another may mean that there is a progressive thinning between the two or that there is a fault with the thickness change all taking place at the fault. Seismic sections are continuous, so the changes between distant points are visible in between. Whether a change is abrupt or progressive can be seen on the section if the change is visible on it at all.

Lithologic tops like "top of sand" are correlated from well to well. They are significant in terms of rock types that may be encountered in a well, but they do not necessarily have meaning in terms of geologic time. Sands may occur in more recent sediments in one well than in another. So representing a "top of sand" pick by a line drawn between wells may have no meaning, except that the highest sands in the two wells were picked. The line may cut across depositional surfaces. Seismic reflections tend to parallel depositional surfaces.

In velocity determination, seismic methods have usually been considered inferior to well data. Seismically determined velocities were thought of as less correct and therefore were to be adjusted to fit well data. However, sonic and density logs are subject to error from washout of the hole and from invasion of the formations by drilling mud. Synthetic seismogram studies have shown that the sonic and density logs require careful editing, beyond that normally performed, to make them adequate for making synthetic seismograms. Overall, the errors can be corrected by adjusting to a conventional velocity survey that uses shots at the surface and a geophone down the hole, but the details would remain slightly in error, distorted to fit the correct overall velocities.

LOCATION PROBLEMS

Seismic interpretation on a map is dependent on correctness of the map. In general, interpreters tend to accept the locations of shot points and other features on the map without question, simply because they have no choice. It's a matter of using that map or doing without. Also, locations are so very important to geophysics that one tends to think that surely someone saw to it that they are correct.

Well, it ain't necessarily so.

FALLIBILITY OF SURVEYING

On land, surveyors are just people, like recorders, drillers, you and me. They can make mistakes, be lazy, not be very experienced, can have defective equipment or untrained assistants.

Offshore, surveying is almost always done by some kind of measurement of radio signals from some other station or stations. A station can be in space and moving in orbit or on a known fixed point on land. Satellites don't pass by at all the times you'd like them to, so a moving seismic boat must use some kind of less accurate fill-in data between passes. The stations on land can be mislocated or moved. Weather can affect the radio waves.

And after the surveying, the map can be drafted incorrectly. A draftsman can misread a shot point number and carry that error on down the line, numbering at uniform intervals. Or the draftsman can daydream and put down the wrong line number.

An area that has had a number of seismic programs shot through the years has had a mix of surveyors, surveying methods, known points as takeoff. The different vintages, even if each is consistent within itself, may not be compatible. One may be skewed slightly, another stretched. Fitting several sets of surveying together is an enormously tedious, painstaking process, full of unsatisfactory compromises, never as good as it should be. And the work is thankless, as the best of results yield only what people expected all along.

Now I've gone and scared you. I'm scared too, and intend to stay scared. This does not mean that surveyors in general aren't competent. They are. It's just that the failings, when they do occur, are so hard to detect and matter so much. There are a few things that can be done.

First, if you have any control over field operations, postplotting, etc., be aware of the problems and give the people involved all the support, equipment, encouragement they need. If something that would help seems expensive, remember the cost of finding and correcting errors later and, in the extreme case, the cost of a well's failing to find oil because it was drilled in the wrong place.

Another thing to do is to pay attention to the surveying and its problems. Don't assume that somehow the problems will be resolved or go away. The problems may continue to trouble everyone who works on the data for many years. And as interpreters replace each other, the new one usually has to learn that the problems exist and then try to figure out something to do about them.

The usual interpretation situation is one in which the field operations are finished, some of them long ago, and you have to work with what is on the map. How can you detect problems and what can you do about them?

In a simple case, let's say there is a strong, easily recognized reflection on the sections. If two seismic lines intersect, that horizon must be at the same reflection time on both. If it isn't, there is a problem. Provided the reflection is unmistakably the same on both sections, it isn't possible for it to be at different levels on the two lines, at the same place. It there isn't some other problem, like some correction's having been applied to one section and not to the other, then the location of at least one of the sections must be incorrect.

If the horizon is flat on both sections, it will match no matter where the sections are put together, so there won't be any indication of a mislocation. But, with dip, and especially steep dip, mislocation will usually make the section fail to tie. The steeper the dip, the greater the mis-tie.

You are interpreting sections. You have folded one at the indicated intersection and laid it over the other, where the intersection is marked on it. This good horizon, that you can spot a mile away, doesn't meet. You can "hunt," by sliding the folded section back and forth. If the horizons match well at some other location on the bottom section, that may be the real intersection. This assumes that the location error is in a direction along one of the lines and that only one line is out of place. Small errors are more common than large ones, so the farther the section has to be moved, the less likely it becomes that the error is that great. Switch sections, folding and sliding the other one. If this brings about a tie without so much slide, it is probably nearer the truth.

To keep the explanation manageable, I mentioned only one horizon. Of course, though, you look at other reflections too. If the tie is correct, they will all match—if the sections are correlatable, and they usually will be if they are from the same batch of shooting and processing.

If there is steeper dip on one of the sections, then sliding the other across it will produce a near-match with less movement. That isn't necessarily more correct, you can't be completely correct, but this is a way to get a fairly good answer.

Even when you find a match that gives a perfect tie all down the section, the tie doesn't tell you which of the two lines is mislocated. To learn that, you need to check a number of intersections. Then a series of mis-ties along one line indicates that it is the line to be moved.

The line you have somehow decided is out of place, then, needs to be repositioned on the map. This kind of check of the surveying isn't at all precise. You may not want to have the line erased and drafted in the new location on just your evidence. You know what the problem is. Maybe later the survey notes and data can be checked by a surveying specialist, who may be able to place the line where it should be. On your work map, it may be best to just indicate the new positions of shot points by hand. Put your interpreted data at the new locations where, even though you may not have moved the points quite correctly, the data will agree better with other data nearby. The data will agree better because that's what you did—moved the points so as to make them agree. You know the locations were wrong before and now they may be nearer right. That's a big gain.

There are other ways to attack the problem. A fault trace usually appears on a seismic map as a straight or gently curving line or band. If a fault, especially a fairly evident one, zigzags, that may mean that some of the seismic lines need to be "pushed back in line." Work on this kind of problem in Exercise 13–1.

Contouring can also expose a mislocation problem. A friend of mine was interpreting a feature. The records were good, but he had tie problems. With all the data on the map, he couldn't contour it smoothly. Contours wiggled back and forth. Eventually, he separated the data into two groups—the lines shot before a certain date and those shot after that date. Each group could be contoured smoothly to form a nice round feature. And if he moved one group a rather startling distance, about 500 m, the two sets of data agreed nicely and could be contoured smoothly. This was in an offshore area. Maybe one of the radio-surveying base stations had been moved between vintages. Which group should be corrected? Who knows? The best thing then might be to move each group some distance to a compromise location.

In cases in which there is no way to determine which line to move or which direction to move it, you can resort to a rather desperate expedient—omit the data from one line in the vicinity of the intersection. You haven't then wrongly moved a whole line, and you have made it possible to contour the remaining points smoothly. It's not elegant, but you do what you can. Granted, you could have thrown out both lines, but then you might not have much data left to use in looking for oil.

All of this location-checking and adjusting by correlation, you might say, gives another meaning to the term "seismic survey."

FAULTS AT INTERSECTIONS

If the locations of seismic lines are not known exactly, a fault near a line intersection can be confusing. You may see a fault on each of the intersecting lines and plot it on a map, as in Illustration 13–1a.

You can draw a single, straight-line fault through those two, with the throw in one direction—in this case throw is down to the southeast, and the fault trace runs NE–SW.

In Illustration 13–1b, let's look at another pair of fault indications at an intersection. How do you draw a fault through those two? Which way is the throw? All right, I don't think it can be done, either.

You can be fairly sure that there didn't just happen to be two faults near that intersection, with each so nearly perpendicular to its line that it doesn't cross the other line. That would be asking a lot from coincidence.

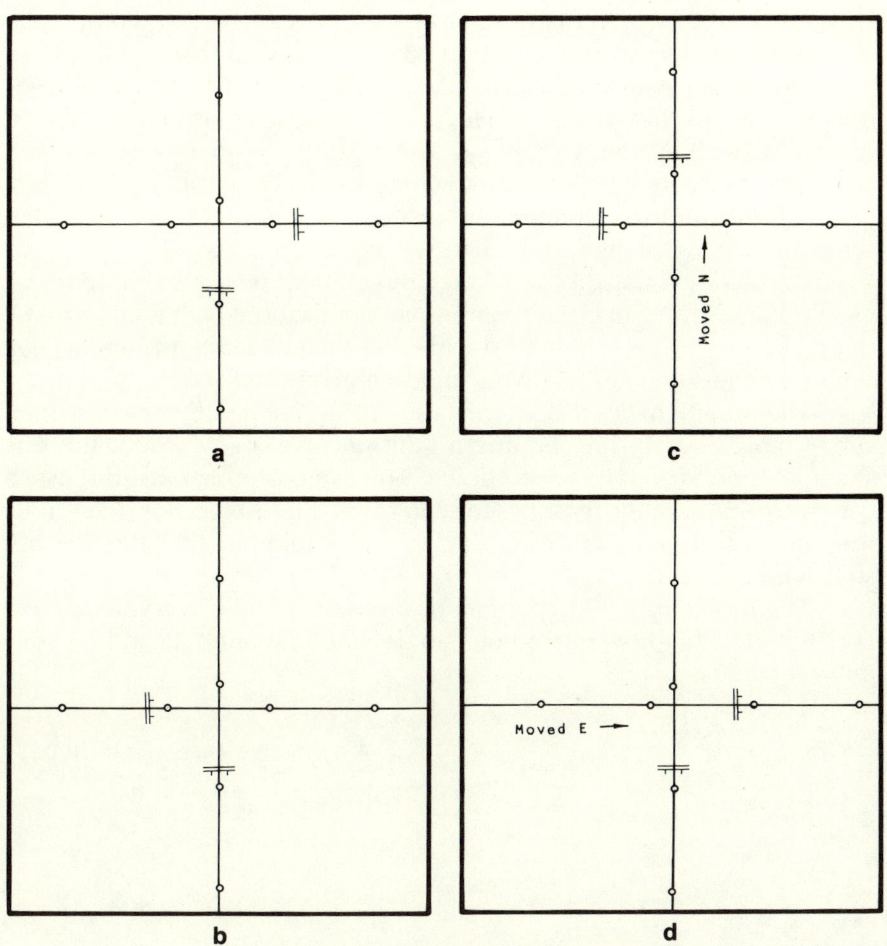

Illustration 13–1 Fault at intersection

At the start of this topic I cast doubt on the line locations. Let's try moving a line a bit, to see if a slightly different location might solve the problem. Maybe the N–S line is too far south. Let's move it northward enough to put the fault on the other side of the intersection. We get the situation in Illustration 13–1c. Look at it.

Moving the line northward makes it easy. The fault runs NE–SW and is downthrown to the southeast. Of course, the direction isn't necessarily exactly NE–SW; that depends on how far we moved the line. But it gives a general idea of direction.

Maybe that line was in its correct position though, and the other one should have been moved instead. Well, let's try that. We move the E–W line to the east enough to put its fault indication on the other side of the intersection, as in Illustration 13–1d. See what that does.

Easy again. Moving the line eastward made the fault run NE–SW, downthrown to the southeast. Hey, that's what we said before! Also, if we had moved the lines sideways instead of along their lengths, we'd again get the same answer. Don't accept my word, try it.

It turns out that assuming the problem is one of location gives the same answer, no matter which line we move. That's good.

Granted, we only know a rough direction of the fault, but that's all we have anyway, from close-together fault indications. What this general direction does for us is help us decide that the two fault indications may represent the same fault so we should connect them.

The situation described can also occur for another reason than survey problems. In interpreting a fault on a seismic section, there is some latitude for interpretive choice. The same problem just discussed can occur for another reason—the fault was interpreted not quite in its true location. To handle this correctly, you would not shift the line but shift the fault on the line.

The problems in this chapter are some of the ones you will be most concerned with—how much you can depend on your data and on your interpretations.

CHAPTER
14

Side
Issues

We have taken up the main things to be done as direct parts of seismic interpretation. Now we'll go into some other things that you need to know or be aware of to make your interpreting go smoothly and successfully.

WORKING WITH PEOPLE

The main parts of interpreting—picking reflections and contouring maps—are things you do yourself. You may occasionally ask someone's opinion of a pick or a contour, but even that discussion will probably be brief, which leaves you back to solving the problems alone. A good interpreter is one who can work well alone, trying different approaches and resolving problems. But to be most effective in the interpreting and the other activities related to it, an interpreter must also be able to work well with other people, communicating with them, getting benefit from their special types of expertise.

WORK ALONE

In doing this private investigation into the nature of some part of the earth, some of your best work may be done when you are uninterrupted. You may sometimes have occasion to work late or to work on a holiday, when there is no one else around and the telephone doesn't ring. The amount of sheer interpretation you accomplish at some of these times may amaze you. You might resolve a difficult line-tie problem that had hampered your interpreting for a long time, or you might find a new and better way to contour a feature. You may wonder what you could accomplish if all your work could be this undisturbed. The special occasions allow your mind to resolve problems that you have been worrying about. The lack of pressure lets you relax and consider things in a different way. But if you had the same undisturbed routine regularly, it could easily become boring. Then you might find that occasional meetings with other people would so stimulate your thinking that you would wonder how much you might accomplish if all your days could be spent working among other specialists. Both being alone and comparing ideas can give you new insights. Plan to take advantage of both. So how about these other specialists?

WORK WITH OTHERS

The people most often involved in an interpreter's work are draftsmen, secretaries, other geophysicists, geologists, immediate superiors, management. In other places in the book we've discussed the relationships with some of them.

DRAFTSMEN

Earlier we discussed at length your relationship with draftsmen. To sum it up, you need their skills to make your work presentable. You should communicate effectively with them so they can know what your needs are and so they can make maps and other displays that fit your interpreting needs.

SECRETARIES

A secretary, like a draftsman, puts results of your interpretation into a neat, legible, presentable form. Your reports must be typed, often from scribbled drafts. The secretary should be warned when a long report is to be made up. If possible, give the report to the secretary far enough ahead of the deadline that intermediate drafts can be typed and proofread. There are many problems that will show up in a report. You are making it for posterity. It must be complete and clear.

If you are working toward a deadline, you may not be able to give your draft of the report to the secretary in advance. You may not know the main conclusions until the interpretation is nearly complete. As with draftsmen, you need to do the best you can, perhaps giving the secretary a draft to start on, with the warning that it may change. If the secretary has a word processor, the changes will cause considerably less trouble than if a typewriter is used.

OTHER GEOPHYSICISTS

Your interpreting is, and almost has to be, your own. Other people may make suggestions, but after they make them, it is up to you to make the loops tie and to make the contours reasonable. The others usually won't have the time or inclination to do the real work of your interpretation. And in some of your interpreting, you may find that a suggestion that had sounded good won't work.

But if there are other geophysicists in your office or geophysicists who visit from other offices of the company, there is a lot of knowledge, skill, experience that you may be able to apply to your work. Talk to them about their work, look at their areas, discuss their problems, and offer suggestions. And, when you have a problem or even when you see an interesting phenomenon in your data, ask for their opinions. One thought may help you solve a problem or may turn up a new prospect. You are each doing your own work, but the exchange of ideas and experience can be helpful.

After you gather new ideas, go back to your private interpreting and be alert for chances to apply the ideas. You may use one immediately. Another may stay in the back of your mind for years and then contribute to some area.

When you do get a useful idea from another person, don't be afraid to admit it. Tell the person. Refer to the source of the idea in discussions with bosses. You are not only giving credit; you are also showing that the idea has support. And the bosses may feel that people who can cooperate technically are useful people.

GEOLOGISTS

Geologists are somewhat in the same relationship to you that other geophysicists are. The geologists are specialists with knowledge and experience different from yours. You and they can profit from an exchange of ideas. They may be working on the same area you are interpreting and may well have ideas that conflict with yours. Good. This is an opportunity to compare ideas and maybe come up with a new concept that neither you nor they had considered before. After the discussions, go back to your area and see how the new ideas work out.

MANAGEMENT

We have also discussed some elements of your relationship with your boss and with higher management. Your boss will give you assignments—what areas to work and the like. You may sometimes receive what seem like foolish orders, for instance, to stop work midway on an interpretation to work on another area. Those orders are usually more reasonable than they seem. Other factors that you aren't aware of influence the decision. A result of a well somewhere can change a company's interests, so your area is no longer one that might be drilled. Don't be unhappy. Work the new area and be glad your interpreting skills are considered worth applying to the new area.

As a technical specialist, one of your main duties to your superiors is to give your technical opinions clearly, so they know what you think and can take that into account in forming judgments. If you have opinions that do not agree with theirs, express those opinions clearly also. If, when they are aware of your opinions, they decide against your ideas, accept their decisions gracefully. You have done your duty and they have done theirs.

SEISMIC CONTRACTORS AND DATA PROCESSORS

Early in a career in interpretation, you may have little association with seismic contractors and data processors. Program planning and field supervision will probably be done by more experienced personnel. This would also apply to choosing the parameters to be used in processing and the checking to see how well the sections show the features.

But later, you will be checking the processing of sections and perhaps even setting parameters. As I mentioned in another chapter, some of your best work with processors can be open discussions about what can and what can't be done about problems. Processors' representatives are almost all technical people. They are quite willing to tell you the undesirable effects of any processing step if you just ask. You can get better processing by applying your and their specialized knowledge together. The best processing I ever had done was also the time when it was easiest for me to get to the processing center for discussions.

As for crew supervision and other work in the field, you may visit the crews to learn about field operations and about the particular problems in the area. But when you are planning a program or when your program is given to a contractor, it can be wise to ask the contractor's opinion of the program. Field supervisors may suggest changes that will save time and money and that will produce better data. Similarly, if you are involved in hiring crews, the supervisors will be able to advise you on the types of equipment that will work best in the area.

WORKING WITH THOSE **BIG** PIECES OF PAPER

Much of an interpreter's time is spent, used up, wasted, however you want to look at it, on the physical handling of seismic sections and maps. There are a few tricks to make this use of time more efficient.

MOSTLY SECTIONS

Sections often come from the printer rolled. To keep them in an accessible file, they can be kept in a wire map rack on casters. If there are only a few lines in the area, this works fairly well; but some hundreds of sections would require a number of racks. The racks take up expensive office space and call for a good bit of wheeling racks around and poking sections in place. Also, there is a lot of rolling and unrolling to get to the part of the section you want to look at.

Folded sections are easier to file, and they take up less space. Accordion folded, they can be opened to the right place like turning the pages of a book. Some interpreters are bothered by the folds, feeling that the fold distracts the eye in its look along a reflection. Most geophysicists do not object to the creases. By folding sections, or printing them on pre-folded paper, they gain the advantages of ease of filing and page-turning and the convenience of having a stack of sections nearby to work on.

A further step is folding not only accordion-fashion but also in the other direction, to fit in a standard filing cabinet. For groups of a few lines, there are file boxes that can be easily moved about. A slight disadvantage of folding both ways is that the sections tend to wear out at the junctures of the two folds, requiring a lot of application of tape to the backs of sections. But the convenience so far outweighs the small disadvantages that most interpreters keep their sections folded both ways and in filing cabinets.

A problem with all section filing systems is that more than one person may need a certain section at the same time. Uncounted, expensive, man-hours have been spent on—"Do you have Line 34?" "No, maybe Joe has it; he was looking for some sections."—and also on—"Why didn't you put them back in the file?" "I was planning on working the area some more and didn't want to have to dig them out again." You know how it goes. More prints, already made and in the file, would help. Or getting a fresh set of prints to work on an area and not using the "file copy."

When an extra version of all or part of a section is made, it is helpful to tape the film of that extra version to the film of the regular section. This eliminates all sorts of filing, hunting, and not even knowing that things exist. It is frustrating, not to mention inefficient, to complete an interpretation with what you think is all the data, and then, when you are presenting your recommendations, to have someone mention remembering an improved version of one of the critical lines. If the films had been combined, that version would have been included in your print of the section.

When the films are combined it helps to put a notice on the regular section, so people will know to unfold it to find the other version. If SPs 250–705 were migrated, tape that migrated part to the unmigrated film of the section and, on the unmigrated one, put a bracket, or arrows or something, from SP 250 to SP 705, with a note "See Migrated Section" or "This Part Also Migrated" or something. But don't just say "Migrated," or people may think that part itself is migrated.

A way I started doing in large areas to reduce the paper handling is breaking the areas down into smaller ones. To work in the east end of a big area, you normally don't need the west end of a long line. You can divide the area into blocks (preferably along the grain of the lines), cut sections apart, and file the parts of sections by block. Then, if you are working on a prospect in Block 3E, just get out the 3E segments, and maybe 3F if your prospect slops over in that direction. It gets a bit messy when you find yourself working at the intersection of four blocks.

For that situation and for a feature like a prospect or a producing field that gets looked at a lot, make a set of segments to cover it, in addition to the regular blocks. This calls for another set of prints, but to repeat, they are much cheaper than an interpreter's time. If you use these block and prospect files, you should also have a file of full-length sections to refer to for regional dips, overall character of a reflection, etc.

To go a step further, draft the block boundaries and block numbers on the film sections, so prints of sections for a block can be easily made. Also draft the line number in each block. Even make a set of sepia masters of the segments in a block, so a set can be quickly made without looking through the originals and then indicating parts of them to be printed.

This filing by blocks can reduce the problems with rolled sections, as the rolls for a block are short. Or the block sections can be folded together, making a neat packet of the block. Better than either, they can be filed flat with each block in a clip or oversize folder.

A commonly—almost universally—used way of reducing the paper handling is to reduce the scale of the sections. Shoot them down photographically from the normal full scale to half that scale or some other convenient small scale. Prints from the reduced film are easier to handle in whatever way you keep sections. Half scale, which is half the length, is only a quarter of the area, so reproduction cost and filing space are greatly reduced.

Reduced sections can be picked more quickly just because they are easier to toss around. And they give an overall view without requiring an extra-long table or floor space. Your arm is more likely to be long enough to reach a feature you want to mark as you squint down a section. However, some seismic detail is easier to examine on large sections. And large sections make better wall displays for meetings. It is useful to have both scales available.

My own preference is to have reproducibles and prints of:

1. Full-scale, full-length sections for fine detail and for displays to be seen from a distance,

2. Half-scale, full-length sections for regional work,

3. Half-scale segments filed as blocks and as much-worked prospects.

Note that this calls for three to four transparencies of each section and corresponding prints in the file and a rather free making of prints as you go along. This will cost a company some hundreds or thousands of dollars. But that cost is tiny compared with people hunting for sections, waiting on sections, wrestling with paper, not making comparisons because it just takes too much time when there is a deadline—in general not finding oil because they are too busy saving on reproduction costs.

MOSTLY MAPS

Whether you keep your work sections rolled or folded, there are still a lot of large sheets of paper and film to handle in rolls. Maps are often large so they can show enough area at a convenient scale, and sometimes you may want to keep your work maps rolled. You need to learn to handle all these large sheets in a way that keeps your paper prints in good condition to work with and show to people. Especially, you want the film or paper transparencies to continue to be in condition to make good copies when you need them.

Paper prints are usually folded, but the transparencies are necessarily kept in rolls. Folding would damage paper transparencies by causing them to tear along the folds after repeated foldings. It would not harm film transparencies in that way, as they are very tough and could be folded many times without tearing. But with either material, the folds show up as lines that will appear on prints made from them. The transparencies must be kept in rolls.

There are several ways to file rolls. A wire map rack on casters is designed for portable filing of a few rolls. A vertical map cabinet typically has upright tubes in it, pivoted so all the tubes together can be angled out for getting rolls out or putting them in. One type of file holds the maps and sections unrolled and suspended vertically from hooks by special gummed strips attached to the maps. A horizontal map file usually has square pigeonholes to slide the rolls into.

Handling large rolls calls for some special skills. Handling one long sheet in a roll is fairly easy. Just roll or unroll, keeping it rolling straight, so the ends don't spiral out, which would make the final roll longer than the width of the sheet. You should be careful not to cinch it much. Cinching is tightening after it is rolled loosely. Any abrasive dust or dirt will scratch the surface when you tighten the roll. Eventually there would be a number of scratches running in the long dimension. It is better to start the roll with a small diameter and roll tightly around that.

When there are several sheets to be rolled together, the rolling becomes more difficult. With patience you can roll a number of maps or sections together so there are not edges sticking out to get banged up.

Large maps are clumsy to look at for information. If a map is spread out on a desk, there may not be room to look at a second map. In general, we haven't learned the lesson of the comic strip. Each picture in a comic is drawn large—a size that is convenient to draw on. Then, for the newspaper, it is reduced to maybe a quarter of the original scale, to print, handle, look at. The newspapers are printed in much greater numbers than our reports, so economics favors the reduction more for them than for us. But a report that consists of a bound text with a thick pocket in the binder, full of big folded maps, is hard to read.

Once I went to a central office on my return from making an interpretation in another city. Asked the boss what he wanted to do. He said to spread the maps out on the floor and go through the report. Well, in addition to the big folded maps, I had bound with the text a set of copies of the maps, each one reduced to page size. We didn't get around to opening a single big map.

Another time, a group of us had been working for a long time on a very large area. It required a number of large maps to cover the area at a convenient scale to work on. After they were interpreted, they were so big that there was no good way to see the overall picture. In desperation, I had them all reduced and spliced together into a single map. It was huge, too, but could be put on a wall of a large conference room we had. People put it on the wall and looked for regional trends. They chose to work in the large room so they could see it. The reduction was such that we couldn't really read seismic times on it, so we knew we needed the larger scale maps in reserve. But I was never aware of anyone unrolling one of them.

The large-scale maps are needed for the rare occasion when it may be necessary to check things in detail, as when a well location is to be selected or changed or when someone is going to re-interpret the area. The big maps should be available, but for most uses, the small maps are much better.

The highest quality, sharpest, small .maps are photographic reductions made in reproduction shops directly from the big ones. But an office copier that has the ability to make reduced copies is expedient for making small copies quickly.

A further refinement in making reduced maps for reports is to not just reduce but also to simplify. The lesson here is from toys. A toy truck for a child to play with is not a reduced exact copy of a full-size truck. It is a simplified version with the main features represented. A scaled-down truck would have too many detailed projections, which would either break or hurt the child. Similarly, a small-scale map, ideally, would have contours but maybe not all the data. And the contours would be bold.

"MAKE ME A COPY"

People unfamiliar with the various reproduction processes that are available tend to say, "Make me a photocopy of this," or "_____ a copy," "_____ a print," _____ a film," without actually referring to the specific processes. This can lead to the use of a process more expensive than needed or less satisfactory or both and can cause time-consuming reprinting. Even if you don't plan to handle reproduction yourself, you need some idea of the process to ask for so you can discuss your needs with the person who does perform, or order, the reproduction.

We'll take up the main processes and their special characteristics. A major separation of the processes is between those that form positive images and negative images.

These two effects are common in ordinary life as well as in reproduction. Light either darkens or fades things. Ripening darkens some fruit, so a leaf blocking the sun from an apple forms a negative image of the leaf—dark where the sun struck the apple, light where it didn't. Similarly a negative image of a bathing suit is formed on the skin of a sunbather. But clothes fade in the sun, forming a positive image of a patch—light where the sunlight reaches the cloth and dark where it doesn't.

Some of the reproduction processes are like the darkening and fading, but they are faster and have a fixing process. The natural processes do not give permanent images. More sun will continue the process and, if the blocking leaf, bathing suit, patch is removed, will obliterate the image. A fixing step is one that makes the image no longer sensitive to light.

DIAZO

In doing seismic interpretation most of your time will be spent looking at, and leaning your elbows on, diazo prints. For large maps, seismic sections, and geological cross sections the diazo process is the most practical day-to-day reproduction system. Prints and sepias can be made quickly at a reproduction shop or on a diazo printer in the office.

Much of the way drafting is handled is determined by the requirements of the diazo system. Maps are drafted on transparent paper or film, so diazo sepias and prints can be made from them. India ink is used to make the lines dense and highly opaque to reproduce well by diazo. Seismic sections are made by processors, photographically, on transparent film, to be copied on diazo prints. Map sizes are usually kept within the limits of most diazo printers, 42 inches wide. The printing process is continuous, so the map can be quite long.

Diazo prints are called by various other names—ammonia print (referring to the developer), dyeline (referring to the diazonium salt as a dye), or specific names for the kind of diazo print—blueline print, blackline print, sepia. Blackline prints on opaque paper are the most familiar to interpreters. Sepias have brown lines on transparent paper or film. A sepia can be used as a master to make other diazo copies from.

The diazo coatings fade rapidly in intense light, but can combine chemically with ammonia to form the permanent dark colors. In practical use the transparent original and the diazo paper are together fed into a machine in which they are carried around a rotating glass cylinder with a light source inside it. Light exposes the sensitized material, and ammonia fumes develop it.

Some ammonia apparently remains in the diazo print for a long time. Ammonia is a bleach, so you should not leave a valued photograph or colored magazine page in contact with an ammonia print for many days.

There are subtleties in using diazo. A transparency of shot point locations is usually kept as a base map. Then a sepia made from it can have additional data drafted on it, for instance, the seismic times for a certain horizon. Prints can then be made from either transparency. Another horizon can be put on another sepia made from the base map. Then there may be occasion to make another sepia from one of those sepias, maybe to add contours to, so prints of it can be put in a report or used by several exploration people. Each stage of copying from an earlier stage is called a generation. Some clarity is lost in each step, so copies from copies are not quite as sharp as earlier generations. Keeping the number of generations small is desirable.

One reason for loss of sharpness is the diffusion of light through the thickness of the transparency in the printing process. If the original is turned face down so the image on the original and the emulsion touch, the print is backward. This principle is often used when a sepia is made. The original is put face down, producing a "reversed sepia." The sepia is transparent enough to read from the other side. When a print is made from the sepia, the sepia is placed with its image surface down, touching the emulsion of the printing paper.

Reproduction specialists, like other people, want to be appreciated for doing good work. They make prints that will please their customers, prints with good sharp lines on a nice white background whenever possible. When the master they have to print from is poor, they may not be able to achieve both. Customers notice a dirty-looking background immediately and criticize or, worse, take their business to another printer. This makes the printers tend to lean toward the white background and sacrifice legibility a bit.

That is usually satisfactory for prints, but sepias, like other intermediates, are another story. A little background color doesn't hurt a sepia. When a print is made from the sepia, the exposure can be set to give the print a white background. But legibility lost by making the sepia too light cannot be recovered. This is where customer-pleasing gets into the act. The printers may know that a slightly darker sepia is better. They also know it may lose them the customer. So they tend to make it, too, without background. You can get better work from them by letting them know you will accept a little background in sepias to keep details from being lost. But, if they don't quite understand, they may go too far and give you a too-dark sepia. That isn't good either.

These three principles are worth remembering:

1. Don't get many generations from the original,

2. Use reversed sepias when you can,

3. Leave a little background in the sepias.

They will make your interpretations look better.

BLUEPRINTS

Blueprints are not diazo prints.

Blueprinting is an old process that forms a negative image on paper in white lines on a blue background. Blueprints were used for house plans, to such an extent that house plans are often called blueprints. As far as I know, the process is not now used in oil exploration. The main reason for bringing up blueprints is to point out that they are not the same as blue**line** prints.

PHOTOGRAPHY

The main purpose of photographic film transparencies is to make diazo transparencies or prints from them. Their very opaque lines and very clear backgrounds make it possible to make the best prints from them. But their cost is prohibitive for really general use. They are used by data processors to make seismic sections and are used by companies for some maps, especially in enlarging or reducing maps.

A photographic image can be formed in two ways, by contact or by projection. In the contact method, a transparent sheet of paper or film with an opaque image on it is laid over the sensitized material in tight contact. In projection, an image is focused by a lens onto the sensitized material, as in an ordinary camera.

Projected images are used to copy maps at different scales, usually smaller than the original. Reduction usually makes the drafting of the map look better, by reducing the size of any irregularities or bobbles. But too much reduction makes the lines too fine to show up well. Enlargement makes the drafting look crude and the numbers oversize. A photographically enlarged map is usually used only as a base to draft a new map from, so the lettering is again a good size but on a larger-scale map.

A different use of photographic copying is on microfilm. This is ordinary projection copying, but at a great reduction, and normally highly automated. The film is specified by width, with 16mm and 35mm films in common use for small documents like letters and computer printouts, and 35, 70, and 105mm widths are used for large maps. The microfilm is negative, but this is no great disadvantage. It can be inspected in a reader. In that use, the mostly dark negative image is easier on the eyes than the glare of light projected through mostly clear film.

The main idea of storing maps on microfilm, though, is to be able to locate and reproduce them readily. Rolls of film can be cut apart into separate images, which can be easily filed. In the case of 105mm, positive prints the same size as the negatives can be handed out to the people who may sometimes need large copies of the maps. These small prints are easy to file, just by dropping them into a desk drawer or keeping them in a box. Then, when a map to work on is needed, it can be called for by title or some number code or even by showing the paper print to the person in charge of the files. The cost of making an enlargement from the microfilm is greater than the cost of making a diazo print from a sepia, but storing and finding the microfilm are much cheaper than paying office rent on space for many large sepias and having people waste interpreting time unrolling them hunting for the one wanted.

MONITOR RECORDS

The monitor records made in the field to check on the recording quality are of two types. One is xerographic, like an office copier. The other kind is made on light-sensitive paper that is also developed by light. There is no fixing process, so the paper remains sensitive to light.

You can recognize these still light-sensitive monitors by their brown lines on a lighter brown background. If you have some of them in the office, keep them away from light and air. If you have occasion to get them out, go ahead and use them freely but don't leave them lying about in the room for days. And if you need to copy them, try the copying method first on an expendable corner of a monitor before exposing them to the light of the copier.

XEROGRAPHY

Office copiers are made, and used, for copying letters, reports, and the like. But in addition to those simple uses, a copier can help you considerably in your interpreting. Spend some time with it, learning its ways. The convenience of having it in the office is one of its main advantages. Another is its ability to copy from opaque originals, rather than the specially prepared transparent ones necessary for diazo. The process used in most of the copiers, xerography, is the one we will consider.

If you have only a print of a section but no transparency, you can make another section to work on with the office copier. Then you don't have to mark the original and can keep it for a master. Or even if you do have a transparency, you can often save time by copying from prints rather than waiting for a diazo copy. You will have to copy the section in segments and splice them together. If the copier will reduce, you can get a smaller scale section to work on.

You can get a more or less unmarked section from a print that already has colored picks on it. The copy is in black and white, and some of the colors do not reproduce well, so the traces of the section are again apparent. But some colors copy so well that the section is blacked out where they are. Working on the copy is often easier and less confusing than trying to make an independent interpretation on an already colored section.

Geophysics is a technical field, so at some time you may have some map or something with exact tolerances to copy, and decide to use the handy office copier, then learn to your dismay that the proportions aren't quite right on the copy. It may be slightly larger or smaller than the original, or one side of the image may be longer than the opposite one. The projection allows things to get slightly out of adjustment. For copies of typed letters or notes this slight misadjustment isn't important, so the machines for office copying aren't intended for exact-scale copying.

Other xerographic equipment has been designed for these close tolerances. It can make copies of maps, seismic sections, etc., at true scale, or any of a number of reduced scales. These machines are much larger and more expensive than office copiers. They are found in reproduction shops, where they can be used for many customers. Using them calls for sending the original to the reproduction shop. If the shop picks up and delivers, it may take about a day to get the copy, or you can take the original, wait a short time, and bring the copies back.

Xerographic copies are supposed to be very permanent. They are, in many uses. But vinyl plastic, like many notebook covers, tends to attract the plastic powder image from the copy. If you leave a copy next to the cover of a notebook, you may find that both notebook and copy are ruined, each having parts of the image.

After all these peripheral matters, let's go back to the subject of interpretation. Turn to Exercise 14–1 and get some more interpreting experience.

CHAPTER 15

Fun and Games

We have gone over the work of interpreting and other day-to-day duties that you may have as an interpreter. There are some other aspects of the business to be considered. You are likely to have a lot of fun, both in doing the work and in side benefits of working in oil exploration. And there are a number of technical advances in interpretation that will reduce the routine, make your work more effective in finding oil, and give you the stimulation of learning new things.

BEING AN INTERPRETER

In addition to the job satisfaction available in many kinds of work, there are two special sources of pleasure and excitement in a career as a seismic interpreter. These are the fun of being associated with different parts of the world and the excitement of having wells drilled on your recommendations and waiting to see if they produce.

The first of these can happen to you in different ways. You may be transferred and so have the opportunity to move to a different location in your country or to a different country. This can be a wonderful experience. You can have first-hand experience of geologically different localities, different cultures, different accents or languages. People in seismic exploration are commonly world travelers. And the traveling in this way is far superior to being a tourist. Unlike the tourist, you get to live in the places rather than just passing through, you work with the local people in the office, learn to conform to their ways. And even if you don't do much actual moving from place to place, you are likely to work with seismic data from many different regions and to associate with people who have traveled a lot.

The second source of fun, the chance of finding oil, is enhanced by the fact that you are likely to be wrong most of the time. More often than not, when a well is drilled on your recommendation, it will be dry, or it will have some oil or gas but not enough to pay for the effort of producing it. The knowledge that you will frequently fail adds to the excitement of anticipation and to your delight when you do find a commercial field. Activities that you can do successfully almost every time become routine and lack excitement.

When oil is found, you will notice how many people point out their part in making the discovery. Many of them are right. You didn't discover it single-handed. A geologist may have found reasons to shoot the area. The seismograph crew shot it effectively, and processors enhanced the reflections to help the interpretation. Your boss assigned it to you and maybe guided your work on it. A series of people on up the chain of command decided it was worth the risk of drilling the well. But you had a big part in all this and so you can be proud of your work.

GO TO THE FIELD

This is a book about interpretation, not field operations. But the two go hand in hand. The field work is done so there will be something to interpret. Interpretation is done on the results of field work and is dependent on type of terrain, shooting method, quality of geophone planting, etc. The data is poor where an obstacle caused some shots to be skipped. A pattern of change in data quality may be related to the hills and valleys or to spread lengths.

Interpreters who have no field experience are at a disadvantage. They are unlikely to understand the header information or notations on the section of special field conditions. And when involved in program planning, they will tend to pay insufficient attention to physical obstacles.

If you are, or are going to be, an interpreter, get some field experience whenever you can. Work on a field crew during summer vacation from school. As a new interpreter see if the company will send you to the field for awhile for training. During an interpreting career visit the field when you can, either to supervise or just to learn. Field operations change with conditions and with progress, so you never advance in geophysics beyond the need to see how the field work is being done. Sometimes your knowledge of the interpreting problems will allow you to make a useful suggestion for the field work. Other times, being aware of some condition in the field may help you with your interpretation.

Maybe you've seen a radical change on several sections and think it may be a fault. But in the field you notice that there is a change of surface conditions, the edge of a swamp or something. Maybe the lines cross ridges, and a change of program will put them parallel to the ridges, speeding up the field work and improving the record quality.

A side benefit of visiting the field is the fun and adventure. You see different places and problems. You are out of the office and in the weather. Working in the field, you can become exhausted but go in at the end of the day feeling like you have had a holiday.

USING THE NEW TECHNIQUES

Seismic exploration is blossoming with new techniques that will change the way it is done, making the routine work faster and allowing more subtle traps to be found. Some of these techniques are already in use but are not yet widespread in the industry. Others have been made to work but are still being made practical. And others are still being developed.

INTERACTIVE TERMINAL

An interactive computer terminal for seismic interpretation is a video screen like a television set, a keyboard like on a typewriter, and a digitizing board. There may be two video screens. The terminal is connected by wire to a large computer. Processed seismic tapes and map location data are available to the computer. Instead of getting out paper prints of the sections and a paper work map, you see the sections and map on the screens, in color. To pick a horizon on a section, "mark" your pick with the digitizer by watching the screen as you move a "pen" that is part of the digitizer. To tie two sections, you can have them both on one screen, meeting at the intersection. They will automatically be at the same scale and will look like one was folded and placed over the other.

The computer can time the picks indicated by your use of the digitizer. The two screens can be used as you wish, with sections on both or with a map or tabulation on one. The times at shot points can be automatically plotted on the map. A time interval can be calculated by the computer and plotted on a display of the map. All these times are added to the computer's memory. Working with the terminal is very handy and can make interpretation both faster and better. You don't have any poor prints of sections or stretched paper prints of sections.

I've been telling you how to interpret paper sections, and now there are these interactive terminals that people are using for interpretation, enabling them to do the work much better. Never mind, you still need to know how to work the paper sections. Terminals are expensive, large, delicate, dependent on a steady power supply and a computer. Even if you are working at a terminal right now, there will be many times when you will be in an outlying office that doesn't have one, or times when the power is off or when the terminal is down and you are waiting for repairs. Some day, seismic interactive computers may be notebook-size, battery-powered, rugged, stand-alone units, so cheap you can carry a spare. Then you won't need to know how to work paper sections. But that is some years away.

THREE-D DATA

Three-D shooting is very expensive. However, it is so effective that its use is becoming more and more common. The data from it can be assembled into many regular 2-D sections, from lines very close together. These sections can have the data migrated in three dimensions, so you don't have to worry about reflections being out of the plane of the section. To the extent that the migration was done properly, with correct velocities, the sections show only reflections from directly below the line of shooting. These sections are 2-D sections, the best 2-D sections you ever had to work with. The lines are close enough together to show you which way to connect the faults from line to line. You can then interpret the data just like you had been interpreting conventional sections, but more easily and with more reliable results.

There is more in the 3-D data than just super 2-D data. In addition to the parallel sections, you can have a section plotted from a hypothetical line in any direction. Traces can be selected from different lines to make up a line that goes in some other direction, for instance to connect wells with a direct line between them. And in addition to the vertical sections, you can have horizontal sections. A vertical section represents a vertical plane through the subsurface. A horizontal section, also called a time slice, represents just that, a horizontal slice through the earth, at a certain seismic time. A time slice looks almost like a contoured map. The variable-area dark and light bands, seen from above, are arranged somewhat like contours; for instance, a round anticline has a series of rings of alternating dark and light. But the bands aren't contours. Their meaning is more like that of a map of surface geology, where structure is revealed by bands of outcrops of different formations. The time slice, of course, is at a certain time, rather than a certain level, so it is distorted a bit by velocity effects. A fault, seen on a time slice, looks about like a fault on an interpreted map. You see a fault trace, with the bands broken, not meeting from one side to another of the fault.

To interpret by time slices, take a print of one slice and draw a line along one band of color on it. That line is a contour of a horizon at that time. Choose the slice to be at the time of one of the contours you want on your map. Plot or trace the band onto the map. Then for the next contour, put that slice aside and get a slice that is one contour away from that first one, maybe 20 or 100 milliseconds, depending on the dip of the area. Draw this next contour of the same horizon on it. Put the contour on the map. And so on.

Another way to view time slices, to get a quick overall view or just to have fun, is to have each one on a frame of movie film and project the movie. You feel like you're plunging into the earth!

There are mechanical aids for working 3-D data, devices that display sections or slices in various ways. An interactive terminal is particularly good at this. It can even show you, in perspective, an intersection of three sections, two of them vertical and one horizontal.

THE NEW WAVES

When the earth is shaken, several well-known types of vibration are initiated. For all the history of seismic exploration, only one of these, the compressional or P-wave, has been reflected and refracted to give information about the subsurface. Now the S-wave, or shear wave, is beginning to be introduced into practical exploration.

In compressional waves, the motion of the particles is back and forth in the direction the wave is traveling, each particle pushing the next and then bouncing back from it. Shear waves transmit energy by the particles moving from side to side, each particle causing the next ahead of it to move from side to side also.

Shear-wave energy is generated by giving the earth a sideways force, for instance by hitting a stake on the side. To record this energy, a geophone sensitive to side-to-side motion is necessary, one in which the moving mass is free to travel in that direction.

The two kinds of waves behave in different ways, traveling at different velocities and refracting at different angles. By generating and recording both, additional information is obtained. The different paths and velocities, processed and interpreted in their relationships to each other, will surely yield much more specific information about the subsurface than we now obtain.

In another new technique, using the conventional compressional waves, a form of seismic lithology works on the separate traces that make up a CDP trace, by analyzing the differences between them for clues to rock types. The difference in the geometry of the traces of a CDP gather is that, although they are all reflected from the same point, they have different offsets, different distances between source and receiver. Energy reflected from a layer of a certain type of rock, maybe with a certain fluid content, may produce one reflection shape if the energy went nearly vertically down and up but a different character if the sound went at a lower angle, reflecting from it steeply, striking it a more glancing blow. Playing out these traces side by side, instead of combining them into a CDP stacked trace, allows for comparison of the different amplitudes of reflection from the same formation, at the same point. A certain lithology may form a reflection that increases in amplitude with distance from source to geophone, while a different rock may have a reflection that decreases in amplitude with distance. This is a powerful new tool for determining the lithologies in the subsurface.

Both of these are whole methods of investigation, with their own chances of error and things that must be done right. The possibilities are tremendous. Imagine being able to do some special shooting, or maybe just special processing, to get evidence of the type of rock and its fluids.

SEISMIC EXPLORATION GAMES

One of the main factors in making a geophysicist successful at finding oil is experience. The more situations that have been encountered in interpreting, the bigger the backlog to be drawn on. This experience is normally gained from years of interpreting in different areas with data from different shooting and processing techniques. But some of this experience can also be obtained by deliberately "playing" with data—seeing how good your judgment and skill are in artificial, therefore checkable, situations. You can't get all the variety of experience this way, but you can hone some of your skills and then be in a position to get more benefit from the other things that occur through the years. These games are advanced variations of the two earlier exercises in which your maps were grids.

TWO GAMES

I'll present some exercises in the form of exploration games. We'll start with an assumed subsurface configuration and distribution of hydrocarbons. Then, without seeing the "real" subsurface, you can start with a blank map of the area, decide where to shoot, get data from your lines, contour it, decide on well locations, and see if your wells produce. This is almost the complete seismic interpretation process. The main thing missing is the sections to be interpreted. It would be hard to have all lines you might possibly shoot already available as sections. It would also fill another book.

There are two games, representing two different terrain and near-surface situations. In each, assume you are working for the company that has the area leased. The first game is in an old area, one that others have explored unsuccessfully. Their work was done years ago, and you have present-day exploration techniques available to you. You may be able to get better data. You know some things from old reports. There is a river valley without much water, one that trucks can cross easily. But the river valley is so sandy that, even with the best methods, you won't be able to get data. There is a town beside the river. The townspeople are familiar with seismic crews and are disappointed that no oil has been found there. So they now won't permit any shooting within the town limits. A corner of the area extends part way up a mountain. It isn't practical to hire a crew with portable equipment for this small part of the area, and the truck-mounted crew you will be using can't get around on the mountain. There is, though, some outcrop information from stream cuts on the mountain. There are three measured dips to the northwest and one doubtful one to the east. They may be helpful in planning a program.

Management is only willing to provide budget for limited amounts of shooting. They will let you shoot 50 km at first as reconnaissance and to see if you can get information there. If you do get data and have any hint at all of a lead, they will approve another 50 km. Then, only if you have a fairly good lead, they will agree to another 20 km for detail.

In the game, you will plan the reconnaissance lines. We'll have to skip over the discussions in the company before the lines are approved as you laid them out. You're lucky. Your program might have been changed beyond recognition. We also skip your obtaining a crew. You may have had to ask for bids from several companies. The crew shot your lines with some problems, maybe a number of telephone calls to report slow progress or to say a planned line couldn't be shot. We'll assume you studied the terrain carefully before assigning the lines and that no unexpected obstacles cropped up during shooting. Then the tapes were sent to a data processor. There was more communication about problems. Then you picked the sections, having difficulty with some of the loop ties. All this takes place at each stage of shooting.

We pick up the sequence again with your plotting the data on a map, contouring it, and deciding what next step to recommend.

Go ahead now and play Game 1, in Exercise 15–1.

I'll leave you alone with this one now. There is more playing the game you can do. You can develop the field with more wells, finding in the process how correct your field outline was. You can shoot more, and drill more, to try to find other fields in the area.

If you really want to get realistic, you can assume costs for shooting, drilling, production facilities. Using those costs, you can see how profitable your exploration is. The prices can be obtained from people you know in different departments of your company. The costs differ with environment—offshore, cultivated land, desert, etc. So you can vary the prices for these conditions. There are some lessons in this game, beyond the expected ones of practice at assigning lines, contouring, choosing well locations.

This chapter is Fun and Games, isn't it? Exercise 15–2 is an exploration game like the first one, but it is a new area that hasn't been shot before. There are probably similar field problems, but there aren't any old reports. You don't know in advance where the problems are, so you will have to learn about them by shooting, using up some of your limited control on no-data parts of the area.

This area is highly prospective. You have reasons to believe, from extrapolation from nearby areas, that the area has several kinds of possibilities for finding oil. There are likely to be anticlinal structures, fault traps, reefs. Management is so interested in the area that they will be willing to approve additional funds for shooting, any time you find a lead. But their approvals are in units of 50 km or less. After each 50 km, you will have to do your interpreting and plan a new program. Any time you actually find production, they will approve a 3-D survey of the new field. You can draw the field outline from your contouring and read data at every square within that outline and one square around it. Now start working on Game 2, in Exercise 15–2. Try to find all the oil.

Got all the steps done? Found oil? Now you can see how well you have done. You did better if you:

☐   Didn't shoot too many kilometers of line,

☐   Didn't drill dry holes,

☐   Found oil quickly, that is, with few stages of waiting for more shooting or drilling,

☐   Found a large amount of production.

These are listed in more or less the order of when they are done, but their importance is in the reverse order. Finding oil and gas is the most important; the others lead up to it. It's better if you shot more and found more oil than if you saved on shooting and didn't find much oil.

YOUR OWN GAME

In the games I made up you have to look up the data from a list and discipline yourself not to look at other numbers than the ones on the lines you are shooting. That limits your chance to check your own ability to program lines and contour limited data. One thing you can do about the contouring is to assume you have a 3-D survey over the whole area by putting data from the list in all the squares, as you did back in Exercise 14-1. Then it doesn't matter how much you see in advance, except for looking up where the oil and gas are. To get around some of those problems and get more practice, too, you can make your own game. In Exercise 15-3, blank grid maps and blank forms for listing seismic times are provided. You can create different geological situations and develop your ability to understand them.

In real exploration we never really know the exact configuration of the subsurface, but in your game, you do. The original map you drew is the absolutely correct subsurface for your game. It is absolute because it was the source of the data. So you can check the correctness of contouring by comparing it with the original map. This is particularly useful to show you how the flawed data we have—the seismic inexactness in real data, just one number per square in the game—can mislead us.

In addition to these forms you can use any convenient grids. By constructing more games you can test your ingenuity at devising realistic games and improve some interpreting skills by making them and playing them. You may also be able to think up other types of games that will give you practice with other interpreting skills.

NOW TO WORK

It you've read through the whole book and done all the exercises, you should have had a number of the experiences that would otherwise be gained by hard knocks in exploring for oil. I hope it makes you become a competent oil finder sooner than you would otherwise. You've had chances to make mistakes in exercises that didn't have the fate of wells depending on your correctness. You may have also had some fun. Now, get to work on the real thing. Find a lot of oil.

Index

Decorative seismic sections appear courtesy of the following organizations:

Part A   Geo Seismic Services, Inc.
Part B   Petty-Ray Geophysical, Geosource Inc.
Part C   Geo Seismic Services, Inc.
Part D   Teledyne Exploration